Philosophy and the Interpretation of Quantum Physics

Philosophy and the Interpretation of Quantum Physics

Badis Ydri
Department of Physics, Faculty of Sciences, Annaba University, Annaba, Algeria

IOP Publishing, Bristol, UK

© IOP Publishing Ltd 2021

All rights reserved. No part of this publication may be reproduced, stored in a retrieval system or transmitted in any form or by any means, electronic, mechanical, photocopying, recording or otherwise, without the prior permission of the publisher, or as expressly permitted by law or under terms agreed with the appropriate rights organization. Multiple copying is permitted in accordance with the terms of licences issued by the Copyright Licensing Agency, the Copyright Clearance Centre and other reproduction rights organizations.

Permission to make use of IOP Publishing content other than as set out above may be sought at permissions@ioppublishing.org.

Badis Ydri has asserted his right to be identified as the author of this work in accordance with sections 77 and 78 of the Copyright, Designs and Patents Act 1988.

ISBN   978-0-7503-2600-1 (ebook)
ISBN   978-0-7503-2598-1 (print)
ISBN   978-0-7503-2601-8 (myPrint)
ISBN   978-0-7503-2599-8 (mobi)

DOI   10.1088/978-0-7503-2600-1

Version: 20211001

IOP ebooks

British Library Cataloguing-in-Publication Data: A catalogue record for this book is available from the British Library.

Published by IOP Publishing, wholly owned by The Institute of Physics, London

IOP Publishing, Temple Circus, Temple Way, Bristol, BS1 6HG, UK

US Office: IOP Publishing, Inc., 190 North Independence Mall West, Suite 601, Philadelphia, PA 19106, USA

Contents

Author biography		ix
1	**Introduction**	**1-1**
1.1	Introductory remarks	1-1
1.2	Organization of the book	1-3
1.3	References	1-5
	References	1-6
2	**The measurement problem and interpretations of quantum mechanics**	**2-1**
2.1	The wave–particle duality and complementarity principle	2-1
2.2	Copenhagen interpretation	2-4
	2.2.1 The von Neumann processes	2-4
	2.2.2 The quantum Zeno effect and the collapse postulate	2-5
2.3	Entanglement entropy	2-7
	2.3.1 The reduced density matrix	2-7
	2.3.2 Entanglement entropy in quantum mechanics and quantum field theory	2-10
2.4	EPR and Bell's theorem	2-11
	2.4.1 The Einstein, Podolsky, Rosen experiment	2-11
	2.4.2 Theorems of quantum philosophy	2-12
	2.4.3 Bell's theorem	2-13
2.5	Decoherence and the measurement problem	2-17
2.6	The many-worlds formalism	2-18
	2.6.1 The many-worlds formalism and coherent branching	2-18
	2.6.2 Schrödinger's cat and quantum immortality	2-19
	2.6.3 The many-minds interpretation	2-21
2.7	Bohmian mechanics	2-23
	2.7.1 A deterministic non-local theory	2-23
	2.7.2 Beables	2-26
2.8	On observer-participancy or consciousness	2-27
	2.8.1 Wigner's friend experiment	2-27
	2.8.2 The von Neumann–Wigner interpretation	2-30
	2.8.3 The extended Wigner's friend experiment	2-31
	2.8.4 The mind/body problem and quantum Zeno effect: Stapp's theory	2-36

2.9	On observer-determinacy or Orch OR	2-37
	2.9.1 Orchestrated objective reduction	2-37
	2.9.2 Time and free will	2-39
	References	2-40

3 The information loss problem in quantum black holes 3-1

3.1	The Schwarzschild black hole	3-1
	3.1.1 The Schwarzschild black hole and Rindler spacetime	3-1
	3.1.2 Particle motion in Schwarzschild spacetime	3-2
	3.1.3 The Kruskal–Szekeres metric and Penrose diagrams	3-3
3.2	Hawking radiation and the information loss problem	3-4
	3.2.1 Hawking temperature	3-4
	3.2.2 Thermodynamics	3-5
	3.2.3 Gravitational collapse	3-6
	3.2.4 Information loss problem and the laws of physics	3-7
	3.2.5 Unitarity and the Page curve	3-7
3.3	Black hole complementarity	3-8
3.4	The firewall and monogamy	3-9
	3.4.1 Monogamy	3-9
	3.4.2 Firewall	3-9
3.5	ER=EPR and the modification of quantum mechanics	3-10
	3.5.1 ER=EPR	3-10
	3.5.2 Interference as entanglement	3-12
	3.5.3 The extended Wigner's friend experiment revisited	3-12
3.6	The holographic gauge/gravity duality	3-13
	3.6.1 Holography	3-13
	3.6.2 The black p-branes and supersymmetric Yang–Mills gauge theory	3-14
	3.6.3 AdS/CFT correspondence	3-16
3.7	The geometry/entanglement connection	3-18
	3.7.1 Ryu–Takayanagi formula	3-18
	3.7.2 The CFT/BH correspondence	3-19
	3.7.3 The first law of entanglement and Einstein's equations	3-21
3.8	Emergent time from entanglement verified experimentally	3-22
3.9	Other models of quantum gravity	3-24
	3.9.1 Causal dynamical triangulation versus Horava–Lifshitz gravity	3-24
	3.9.2 Matrix models of string theory and noncommutative geometry	3-26

	3.9.3 The measurement problem within noncommutative geometry	3-26
	References	3-27

4 Quantum dualism — 4-1

4.1	The measurement problem as an information loss problem	4-1
4.2	The measurement problem and quantum entanglement	4-3
4.3	Quantum dualism	4-6
	References	4-9

5 Black hole interpretation of quantum mechanics — 5-1

5.1	On the complementarity between Copenhagen and many-worlds observers	5-1
5.2	On the black hole interpretation of quantum mechanics	5-2
5.3	On observers in quantum mechanics and the evaporation of black holes	5-5
	References	5-7

6 On quantum logic and quantum metaphysics — 6-1

6.1	Boolean algebras	6-1
	6.1.1 Definition	6-1
	6.1.2 Boolean algebras as lattices	6-3
	6.1.3 Example 1: The minimal algebra	6-3
	6.1.4 Example 2: Power sets under inclusion	6-4
	6.1.5 Example 3: Classical propositional calculus and Lindenbaum–Tarski algebra	6-5
	6.1.6 Example 4: The logical structure of the classical theory	6-7
6.2	The logical structure of the quantum theory	6-9
	6.2.1 Hilbert space quantum mechanics	6-9
	6.2.2 Experimental propositions	6-11
	6.2.3 The logic of projectors	6-13
6.3	Hilbert lattices	6-16
	6.3.1 Lattice theory	6-16
	6.3.2 Hilbert lattices	6-19
	6.3.3 Hilbert lattice as pasting of blocks	6-20
	6.3.4 Spin one-half system	6-20
	6.3.5 Spin one system	6-23

6.4	Gleason and the Kochen–Specker theorems	6-25
	6.4.1 Gleason theorem	6-25
	6.4.2 Theorems of quantum/experimental metaphysics	6-29
	6.4.3 The set-up of the Kochen–Specker theorem	6-30
	6.4.4 Peres construction	6-32
	6.4.5 More on the assumptions of the Kochen–Specker theorem	6-35
6.5	CHSH quantum game and free will theorem	6-38
	References	6-38

7 Interpretation of the 'Copenhagen interpretation' — 7-1

7.1	Postulates of quantum mechanics	7-1
7.2	From quantum fundamentalism to semi-classical fundamentalism	7-3
7.3	The uncertainty/indeterminacy principle	7-7
7.4	The complementarity principle	7-11
7.5	The correspondence principle	7-12
7.6	The philosophy of Bohr	7-15
7.7	An executive summary	7-17
	References	7-21

8 Neutral monism, perspectivism and quantum dualism — 8-1

8.1	The anthropic principle	8-1
8.2	Quantum dualism as an informational neutral monism	8-4
8.3	The collapse of the wave function and the measurement problem	8-5
8.4	Quantum perspectivism and quantum logic	8-8
8.5	Physicalism, naturalistic dualism and neutral monism	8-10
8.6	Quantum dualism, Wigner's friend and solipsism	8-12
8.7	Simulation and matrix hypotheses	8-15
8.8	The arrow of time and the Boltzmann–Schuetz hypothesis	8-16
8.9	Time, free will and the quantum past	8-19
8.10	The mind/body problem, synchronicity and Bell's theorem	8-19
	8.10.1 The mind/body problem	8-20
	8.10.2 Synchronicity	8-20
	8.10.3 The world-from-decoherence and the world-in-consciousness	8-21
	8.10.4 Bell's theorem revisited	8-22
8.11	Conclusion	8-24
	References	8-24

Author biography

Badis Ydri

Badis Ydri—currently a professor of theoretical particle physics, teaching at the Department of Physics, Badji-Mokhtar Annaba University, Algeria—received his PhD from Syracuse University, New York, USA in 2001 and his Habilitation from Annaba University, Annaba, Algeria in 2011. His doctoral work, titled 'Fuzzy Physics', was supervised by Professor A P Balachandran. Professor Ydri is an Adjunct Professor at the Dublin Institute for Advanced Studies, Dublin, Ireland, and a research associate (regular ICTP associate) at the Abdus Salam Center for Theoretical Physics, Trieste, Italy. His post-doctoral experience comprises a Marie Curie fellowship at Humboldt University Berlin, Germany, and a Hamilton fellowship at the Dublin Institute for Advanced Studies, Ireland. His current research directions include: noncommutative geometry; the gauge/gravity duality; computational physics of string theory; renormalization group and Monte Carlo methods in matrix models and noncommutative field theories; emergent geometry, gravity and cosmology from matrix models; and foundations of quantum mechanics. Other related interests include string theory; quantum information; causal dynamical triangulation; Horava–Lifshitz gravity; supersymmetric and noncommutative standard models; and supersymmetric gauge theory in four dimensions. He has recently published six books. His other intellectual interests include philosophy of physics and existential philosophy.

IOP Publishing

Philosophy and the Interpretation of Quantum Physics
Badis Ydri

Chapter 1

Introduction

1.1 Introductory remarks

Quantum mechanics is perhaps the greatest scientific breakthrough ever achieved. It brought with it a seismic paradigm shift in our way of thinking about nature, and at the same time it underpins most of the dramatic technological innovations of the modern era, as well as providing a profound lasting impact on our metaphysical conception of reality. Thus its value is ontological, epistemic and practical. However, quantum mechanics is only one step (certainly the more colorful, dramatic and puzzling) in a long list of revolutionary paradigm shifts.

The Copernican revolution was the first revolution in physics, in which Copernicus moved away from the Ptolemaic cosmology toward a heliocentric cosmology. This in fact marks the starting point of the scientific revolution. Then comes Newton and his Newtonian physics, which was the first revolution in theoretical physics in which the scientific experimental method was supplemented with the mathematical method. This revolution continued with Euler, Lagrange, Hamilton and others. This classical revolution continued further with the invention/discovery of thermodynamics, which can certainly be reduced to mechanical motion, and electromagnetism, which resisted such a reduction.

This completed (or almost completed) the edifice of classical physics which is truly classical in spirit and methods, i.e. Aristotelian in a precise sense. The Einstein revolution (special relativity and the general theory of relativity) is a part of this classical realm of physics. The relativistic conception of spacetime due to Einstein is mid-way between Newton's absolutism and Leibniz's relationalism.

After Newton came another revolutionary shift—little known to us as physicists—the Copernican revolution in philosophy due to Kant, which (despite its rather complicated nature) can be stated as the fact that cognition (the observer) determines appearances (the world) and not the other way around. Kant called this transcendental idealism. In hindsight this is very reminiscent of one of the most central conclusions of quantum mechanics regarding the irreducible role of the conscious/zombie observer. Yet, according to Kant, metaphysics is not possible.

Then comes the quantum revolution—Bohr, Einstein, Heisenberg, Schrödinger, Dirac, Wigner, Born, Pauli and others in the golden age of modern physics.

To put the problem in a clear perspective we recall first that the philosophy of classical physics is based on the ideas of (i) determinism, (ii) locality, (iii) a world existing without observation, (iv) consciousness being irrelevant to observation, (v) things being knowable and (vi) the world being causally closed. This leads to physicalism, i.e. the view that everything, including minds and consciousness, is reducible to matter.

Quantum mechanics contradicts classical mechanics in every point. So the world which is described faithfully (from experimental evidence) by quantum mechanics is non-deterministic, non-local, it does not seem to exist without observation, the conscious observer by some accounts is implicitly or explicitly crucial to observation, things are not necessarily knowable and the world is not causally closed because of the irreducible character of the consciousness of the observer.

This situation was termed 'quantum mud' by Popper since on one hand there is quantum mechanics and on the other hand there is its interpretation, i.e. the relation between the mathematics of quantum mechanics and the external world is not without a metaphysical burden.

The single most central but controversial aspect of quantum mechanics consists in the so-called measurement problem, i.e. the reduction/collapse of the state vector when subjected to a quantum measurement. The measurement problem is a well posed problem mathematically, involving the physics of entanglement and decoherence. However, the reduction/collapse process is a non-unitary, irreversible and stochastic process which cannot be accounted for by any known physics. It could be caused by the environment, or by the mind of the observer or by some lesser sort of observer-participancy, or it could be caused by the spacetime structure at the Planck scale. Or perhaps there is no altogether non-unitary and irreversible reduction of the state vector but there is instead a unitary and reversible many-worlds formalism.

The measurement problem is therefore the most central question in quantum physics since we are trying to comprehend the world through the lenses of quantum mechanics. Schrödinger summarizes the problem by the intriguing question: Does quantum mechanics provide a fuzzy picture of a clear reality or is the fuzziness in reality itself and quantum mechanics is only providing a clear picture of that fuzziness? Wheeler summarizes the situation by noting that the world according to quantum mechanics can be drawn as an eye looking onto itself.

The quantum revolution underwent another paradigm shifting revolution from within through Bell and his celebrated Bell's theorem with all its physical consequences and metaphysical ramifications.

In summary, the states of the physical system do not really exist before measurement and either the observer determines the world (the measurement promotes potential existence into real existence), which is very similar to Kant's transcendental idealism, or else the world is really a linear superposition of many coherent branches. Thus here, as opposed to Kant's conclusion based on classical physics, quantum metaphysics is a real possibility but tailored with experimental and theoretical quantum physics.

It is well appreciated by string theorists, quantum information theorists and many other quantum physicists that black hole evaporation and the associated information loss problem provide the ideal laboratory to test the fundamental principles of quantum mechanics and their possible modifications by taking into account the equivalence principle.

The information loss problem arises from the fact that a correlated entangled pure state with zero Killing energy (originating in the gravitational collapse of a black hole) when it reaches near the horizon will give rise to a thermal mixed state outside the horizon. The entangled pure state is formed from particle pairs where one particle of each pair is transmitted through the horizon as information loss whereas the other particle is reflected to infinity as Hawking radiation. The existence of the horizon is what makes this problem such a special and peculiar problem.

However, information is thought not to be lost at the end since it will start coming out with the radiation at the Page time when the entanglement entropy between the interior and the exterior becomes maximal (the Page curve). This is the unitarity assumption which will be reinforced in any model based on the holographic gauge/gravity duality.

1.2 Organization of the book

This book is organized as follows.

In chapter 2 (The measurement problem and interpretations of quantum mechanics) we provide a review of the structure of quantum mechanics according to the Copenhagen interpretation and the many-worlds formalism. We also review many physical results/effects and theorems of quantum philosophy with the main emphasis placed on quantum entanglement and Bell's theorem.

The main observation made in this chapter is the fact that the Copenhagen interpretation plays the role of the local view of Reality whereas the many-worlds formalism plays the role of the global view of Reality. This should be contrasted with the situation in general relativity, namely the first-person Copenhagen observers act as local systems of reference whereas the third-person observers of the many-worlds formalism provide the manifold structure of Reality. Thus, Bohr and Heisenberg's Copenhagen interpretation is in no way an approximation of the many-worlds formalism and its intrinsic Nietzschean perspectivism is a physical feature of Reality which cannot be reduced but only complemented with the manifold view of many-worlds or by other approaches.

In chapter 3 (The information loss problem in quantum black holes), toward the goal of drawing systematic parallels between the measurement problem in quantum mechanics and the information loss problem in quantum gravity, we provide a comprehensive review of the information loss problem and various theories of quantum gravity.

Indeed, in this chapter we provide a description of Hawking radiation and the corresponding information loss problem. Then we review briefly the holographic gauge/gravity duality and many other related ideas relevant to the information loss problem and its unitary resolution, such as the connection between spacetime geometry and quantum entanglement.

Starting in chapter 4 (Quantum dualism) we provide a synthesis based on quantum dualism, i.e. the fact that the Copenhagen interpretation provides a local view of Reality whereas the many-worlds formalism provides a manifold view and the two views are complementary not contradictory.

Thus, in this chapter we put forward our first synthesis in which quantum dualism is emphasized over unitarity. In fact, it is believed that quantum dualism itself is unitary via the Copenhagen/many-worlds correspondence, i.e. the complementarity between the local view of Reality provided by the Copenhagen first-person observers and the manifold view of Reality provided by the many-worlds third-person observers.

In the synthesis which will follow in the next chapter we will take the inverse approach and stress unitarity more explicitly than the quantum dualism, whereas in the final chapter of this book we will return to the quantum dualism and relate it to perspectivism and other implicit and explicit characteristics of the Copenhagen interpretation.

In this chapter we also draw systematic parallels between the measurement problem in quantum mechanics and the information loss problem in black holes. Then we proceed to propose a solution of the former along the lines of the solution of the latter, which is based on the holographic gauge/gravity duality. The proposed solution is based on (i) quantum dualism and (ii) the properties of quantum entanglement, in particular its fungibility. This topic is further discussed in the next chapter.

In chapter 5 (The black hole interpretation of quantum mechanics), we explore how the information loss problem and the measurement problem share the fundamental characterization that an initial pure state is evolved into a final mixed state since there is in both cases a part of the system which is inaccessible (the environment in the case of quantum mechanics and behind the horizon for the case of black holes). However, the information loss problem admits in principle a solution via the holographic gauge/gravity duality. The goal is to exploit this formal analogy (and the analogy as we will argue is even physical) in order to extend the proposed solution for the information loss problem to the measurement problem. In this chapter a unitary solution inspired by the holographic gauge/gravity duality of the measurement problem is attempted.

As we will see in chapter 6 (On quantum logic and quantum metaphysics) 'quantum logic' is a theory construed as an interpretation of quantum mechanics, although for some it is a theory which attempts to go beyond. In fact quantum logic may even be understood as nothing more than an interpretation of the 'Copenhagen interpretation'. A much stronger perspective along this line will be taken up in the next chapter.

Boolean algebras, which are relevant to classical logic, are a special case of orthocomplemented modular lattices, which are relevant to quantum logic, where the distributive law in classical logic is replaced with the modular law in quantum logic. The goal in this chapter is to define all these terms carefully. In this chapter we will also discuss in detail the Kochen–Specker theorem which can be viewed as a manifestation of the more fundamental Bell's theorem.

In chapter 7 (Interpretation of the 'Copenhagen interpretation') we take the stronger position that (i) there can be no interpretation of quantum mechanics without the Copenhagen interpretation and (ii) that the Copenhagen interpretation itself requires an interpretation. In particular, it is argued that the complementarity principle is essential to quantum mechanics itself and that as far as the differences between Bohr and Heisenberg go there is a single coherent 'Copenhagen interpretation' which entails both views. In particular the 'classical concepts' for Bohr replace effectively the 'collapse postulate' for Heisenberg. It is the Kantian concept of Bohr's 'classical concepts' which allows us to propose that consciousness is a purely classical concept creating therefore a dichotomy in Nature between a purely 'quantum Reality' and the purely 'classical cognition' of that Reality. This view is termed 'semi-classical fundamentalism' as opposed to the 'quantum fundamentalism' permeating the current understanding of quantum physics.

In chapter 8 (Neutral monism, perspectivism and the quantum dualism) we revisit quantum mechanics in the Wigner–von Neumann interpretation which is the most vocal among all the Copenhagen variants with regard to observer-participancy and/or observer-determinacy. This interpretation, as we will see, is characterized by (i) a quantum dualism between matter and consciousness unified within an informational neutral monism, (ii) a quantum perspectivism which is extended to a complementarity between the Copenhagen interpretation and the many-worlds formalism, (iii) a psychophysical causal closure akin to Leibniz parallelism and (iv) a quantum solipsism, i.e. a reality in which classical states are only potentially existing until a conscious observation is made.

In our view the most important characteristic of the complementarity principle, of the Copenhagen interpretations (such as the Wigner–von Neumann interpretation) and of quantum mechanics itself is 'quantum perspectivism', which bears an intriguingly strong resemblance to Nietzsche's perspectivism (or more precisely Putnam's internal realism). In this chapter we will also discuss, among many other things, two fundamental topics of interest to the philosophy of quantum physics which are (i) the Boltzmann–Schuetz hypothesis and (ii) the mind/body problem.

The main results of this book are summarized in figure 1.1.

1.3 References

A very partial list of landmark results and seminal works which have shaped our understanding of quantum mechanics and guided our interpretational choices made in this book regarding the philosophy and interpretations of quantum physics are included in the bibliography below. I should admit that this list is also a very personal selection.

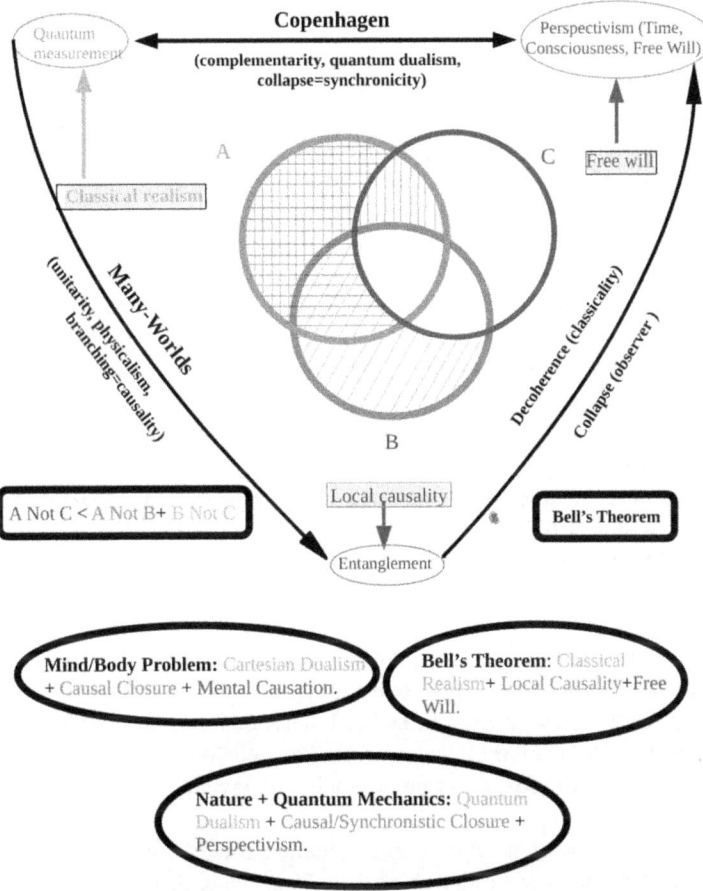

Figure 1.1. Bell's theorem and the mind/body problem.

References

[1] Bohr N 1963 *Essays 1958–62 on Atomic Physics and Human Knowledge* (New York: Wiley)
Bohr N 1949 Discussions with Einstein on epistemological problems in atomic physics *Albert Einstein: Philosopher-Scientist* (Cambridge: Cambridge University Press)

[2] Heisenberg W 2011 *Is a Deterministic Completion of Quantum Mechanics Possible?* transl. Crull E and Bacciagaluppi G

[3] Schrodinger E 1935 Die gegenwrtige Situation in der Quantenmechanik (The present situation in quantum mechanics) *Naturwissenschaften* **23** 807–12

[4] Wigner E 2001 Remarks on the mind body question *Philosophical Reflections and Syntheses. The Collected Works of Eugene Paul Wigner. Part B: Historical, Philosophical, and Socio-Political Papers* ed J Mehra vol B/6 (Berlin: Springer)
Wigner E 1963 The problem of measurement *Am. J. Phys.* **31** 6–15

[5] Einstein A, Podolsky B and Rosen N 1935 Can quantum mechanical description of physical reality be considered complete? *Phys. Rev.* **47** 777

Einstein A and Rosen N 1935 The particle problem in the general theory of relativity *Phys. Rev.* **48** 73

[6] Bell J S 1935 On the problem of hidden variables in quantum mechanics *Rev. Mod. Phys.* **38** 447

Bell J S 1987 *Speakable and Unspeakable in Quantum Mechanics: Collected Papers on Quantum Philosophy* (Cambridge: Cambridge University Press) p 212

Bell J S 1964 On the Einstein–Podolsky–Rosen paradox *Physics* **1** 195

[7] Wheeler J A 1978 The 'past' and the 'delayed-choice' double-slit experiment *Mathematical Foundations of Quantum Theory* ed A R Marlow (New York: Academic) pp 9–48

Wheeler J A 1990 Information, physics, quantum: the search for links *Complexity, Entropy, and the Physics of Information* ed W H Zurek (Redwood City, CA: Addison-Wesley)

[8] DeWitt B S and Graham R N (ed) 1973 *The Many-Worlds Interpretation of Quantum Mechanics* (Princeton, NJ: Princeton University Press)

[9] Zurek W H 1991 Decoherence and the transition from quantum to classical revisited *Phys. Today* **44** 36–44

[10] Tegmark M 1998 The interpretation of quantum mechanics: many worlds or many words? *Fortsch. Phys.* **46** 855–62

[11] Susskind L 2016 Copenhagen vs Everett, teleportation, and ER = EPR *Fortsch. Phys.* **64** 551

Susskind L 2016 ER = EPR, GHZ, and the consistency of quantum measurements *Fortsch. Phys.* **64** 72

[12] Susskind L and Lindesay J 2005 *An Introduction to Black Holes, Information and the String Theory Revolution: The Holographic Universe* (Hackensack, NJ: World Scientific) p 183

[13] Hawking S W 1975 Particle creation by black holes *Commun. Math. Phys.* **43** 199

Hawking S W 1976 Erratum *Commun. Math. Phys.* **46** 206

Hawking S W 1976 Breakdown of predictability in gravitational collapse *Phys. Rev.* D **14** 2460

[14] 't Hooft G 1993 Dimensional reduction in quantum gravity *Conf. Proc.* C **930308** 284

[15] Susskind L 1995 The World as a hologram *J. Math. Phys.* **36** 6377

[16] Maldacena J M 1999 The large N limit of superconformal field theories and supergravity *Int. J. Theor. Phys.* **38** 1113

Maldacena J M 1998 The large N limit of superconformal field theories and supergravity *Adv. Theor. Math. Phys.* **2** 231

[17] Almheiri A, Marolf D, Polchinski J and Sully J 2013 Black holes: complementarity or firewalls? *J. High Energy Phys.* **1302** 062

[18] Maldacena J and Susskind L 2013 Cool horizons for entangled black holes *Fortsch. Phys.* **61** 781

[19] Dicke R H 1961 Dirac's cosmology and Mach's principle *Nature* **192** 440–76

Dicke R H 1957 Gravitation without a principle of equivalence *Rev. Mod. Phys.* **29** 363–76

[20] Boltzmann L 1895 On certain questions of the theory of gases *Nature* **51** 413

[21] Edwards D A 1979 The mathematical foundations of quantum mechanics *Synthese* **42** 1

[22] Nietzsche F W 1964 *The Will to Power* ed W A Kaufmann and R J Hollingdale (New York: Vintage) 267

Nietzsche F W 1974 *The Gay Science; With a Prelude in Rhymes and an Appendix of Songs* ed W A Kaufmann (New York: Random House) §374 p 336

Nietzsche F W 1998 *On the Genealogy of Morality: A Polemic* ed M Clarke and A J Swenswen (Indianapolis, IN: Hackett)

[23] Kulstad M and Carlin L 2013 Leibniz's philosophy of mind *The Stanford Encyclopedia of Philosophy* **Winter 2013**
[24] Chalmers D 2007 The hard problem of consciousness *The Blackwell Companion to Consciousness* ed M Velmans and S Schneider (Oxford: Blackwell)
Chalmers D 2007 Naturalistic dualism *The Blackwell Companion to Consciousness* ed M Velmans and S Schneider (Oxford: Blackwell)
Chalmers D 2010 *The Character of Consciousness* (Oxford: Blackwell)
[25] Jung C and Pauli W 1955 *The Interpretation of Nature and Psyche* (New York: Pantheon)
Jung C and Pauli W 1992 *Atom and Archetype: The Pauli/Jung Letters, 1932–1958* ed C A Meier (Princeton, NJ: Princeton University Press)

IOP Publishing

Philosophy and the Interpretation of Quantum Physics

Badis Ydri

Chapter 2

The measurement problem and interpretations of quantum mechanics

In this chapter we provide a review of the structure of quantum mechanics according to the Copenhagen interpretation and the many-worlds formalism. We also review many physical results/effects and theorems of quantum philosophy with the main emphasis placed on quantum entanglement and Bell's theorem.

The main observation made in this chapter is the fact that the Copenhagen interpretation plays the role of the local view of Reality whereas the many-worlds formalism plays the role of the global view of Reality. This should be contrasted with the situation in general relativity, namely the first-person Copenhagen observers act as local systems of reference whereas the third-person observers of the many-worlds formalism provide the manifold structure of Reality. Thus, Bohr and Heisenberg's Copenhagen interpretation is in no way an approximation of the many-worlds formalism and its intrinsic Nietzschean perspectivism is a physical feature of Reality which cannot be reduced but only complemented with the manifold view of many-worlds or by other approaches.

2.1 The wave–particle duality and complementarity principle

According to Richard Feynmann, the double-slit interference experiment is 'impossible to explain in any classical way', is 'the heart of quantum mechanics' and 'contains the only mystery' [1]. Feynman is also reported to have said that quantum interference is the 'mother of all quantum phenomena' and that because of it 'nobody understands quantum mechanics'.

Quantum interference in the double-slit interference experiment provides the first direct indication of wave–particle duality. Thus the interference pattern is observed even if we send light through the two slits a single photon at a time. It works for photons, electrons and, in principle, for all other particles.

A wave (interference)–particle (path) duality is the first duality in quantum physics. The two descriptions are complementary (not contradictory as in classical mechanics) to each other since they cannot be observed simultaneously.

Another related complementarity principle in quantum physics is the position–momentum duality and the Heisenberg principle. The position x and the momentum p are canonical variables represented by incompatible operators on the Hilbert space satisfying the Dirac commutation relation

$$[x, p] = i\hbar. \tag{2.1}$$

This leads immediately to the uncertainty relation

$$\Delta x \Delta p > \hbar. \tag{2.2}$$

In other words, there is a fundamental limitation on the precision of measurements. The quantum phase space becomes discrete (i.e. pointless!) constituted of elementary Heisenberg cells of volume $\hbar$ containing one state each. The phase space is then fuzzy (since we cannot discern points) or noncommutative (since coordinates are not commuting). This is the prototype for all noncommutative geometry, fuzzy spaces and matrix models, which is one proposal for quantum gravity.

The wave–particle duality is also intimately related to the other complementarity principle: entanglement–decoherence duality. The question then arises: which one is more fundamental?

The evidence seems to point toward entanglement being the most fundamental quantum effect and even interference can be reduced to entanglement, as shown using the ER=EPR conjecture in [2, 3] (the two slits are construed as maximally entangled and as such they are connected via a smooth gravitational bridge).

Wheeler's delayed choice gedanken experiment [4] (which appeared first in his essay 'Law without Law') is perhaps the mother of all interference experiments and is one of the greatest quantum effects, which was verified experimentally in, for example, [5, 6]. This which-way experiment threatens time ordering and causality and according to Wheeler this experiment shows that no phenomenon is a real phenomenon until it is an observed phenomenon.

A source of light sends photons one by one through the paths shown in figure 2.1. The photons pass through a beam splitter BS1. They are either reflected with a probability 1/2 toward the first mirror (path P1) and then to the detector D1. Alternatively, they can be refracted with a probability 1/2 toward the second mirror (path P2) and then to the detector D2.

The paths are determined: if D1 clicks then the path taken is P1 whereas if D2 clicks then the path taken is P2.

We position another beam splitter BS2 between the detectors D1 and D2. Let A be the amplitude of the emitted light. The reflected wave is $i*A$ whereas the refracted wave is $1*A$. We can reach D1 by two routes: reflection + refraction or refraction + reflection, giving the probability amplitude $i*1 + 1*i = 2i$. We can also reach D2 by two routes: reflection + reflection or refraction + refraction giving the probability amplitude $\longrightarrow i*i + 1*1 = 0$.

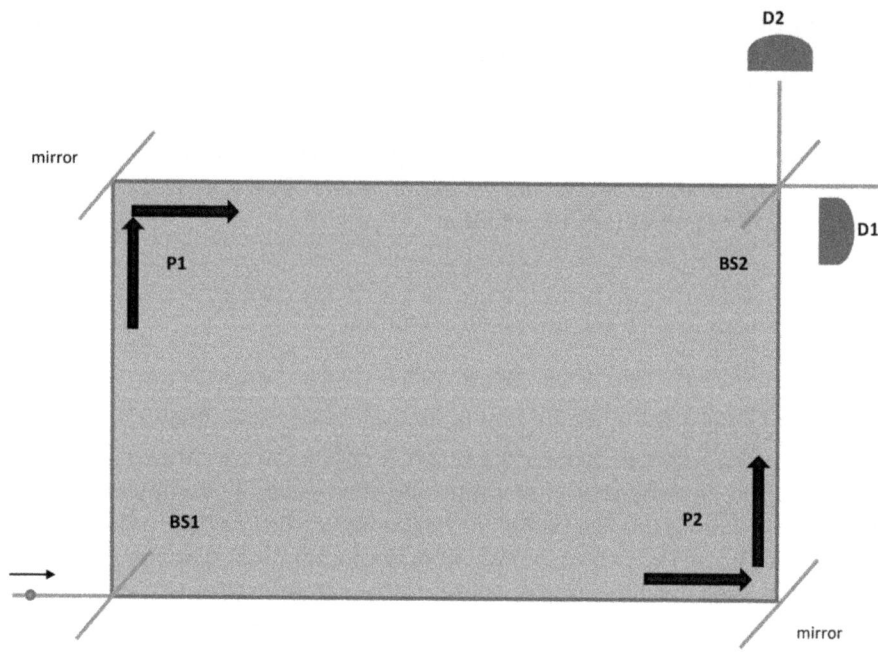

Figure 2.1. The delayed choice experiment.

In other words, the photons reach D1 (constructive interference) always but never D2 (destructive interference), i.e. light behaves as waves when paths are not determined. If only one route is allowed (BS2 removed) light behaves as particles. Whereas if the two routes are allowed (BS2 is not removed) light behaves as waves. This is complementarity.

Wheeler proposed adding BS2 at the last moment after the photons go through BS1 and before they reach the detectors at the intersection points of the two paths P1 and P2. The result does not change: light behaves as waves if we include BS2 and as particles if we remove BS2.

Thus we are deciding retroactively whether the photons act as waves (both paths allowed) or as particles (one path allowed) after they complete their journey. Equivalently, the photons decide to behave as particles (one path) or as waves (two paths) at the last moment although particle behavior requires passing through one path whereas the wave behavior requires passing through two paths.

Wheeler also proposed a cosmic delayed choice experiment. The light source is a quasar whereas the beam splitter is a gravitational lensing. The emitted photons (since millions of years) act as waves by observing interference patterns or as particles if we employ telescopes to determine their paths. We can thus create or alter the distant past by our manner of observing it now, as Wheeler puts it. According to this extreme view even the Big Bang could have been created by our observation.

2.2 Copenhagen interpretation

2.2.1 The von Neumann processes

Quantum mechanics, according to the standard view (the Copenhagen or Bohr's interpretation [7]), is based on two mega-laws (not one) which were mathematically formulated originally by von Neumann in his book [8] (see also [9–11]). These are given by the following.

- *Process II*: The unitary evolution in time generated by a Hamiltonian H given by the Schrödinger equation, namely

$$i\hbar \frac{\partial}{\partial t}|\psi(t)\rangle = H|\psi(t)\rangle. \qquad (2.3)$$

Also one should mention the quantum superposition principles: if ψ_1 and ψ_2 are two solutions of the Schrödinger equation then any linear combination $\alpha\psi_1 + \beta\psi_2$, for any complex numbers α and β, is also a solution. The superposition principle can be given by the path integral.

The Schrödinger equation allows us to compute the state of the physical system $\psi(t)$ at any given time t starting from some initial state $\psi(t_0)$ at the initial time t_0.

- *Process I*: This is the collapse or reduction postulate termed process I by von Neumann (process II is the unitary evolution). This allows us to compute the state of the system when we subject it to a quantum measurement. Explicitly, it states that the state of the system after measurement will collapse to the eigenstate in the Hilbert state corresponding to the eigenvalue determined by the outcome of the measurement process. The collapse postulate should be coupled with Born's statistical rule which determines the probability or frequency of finding the various outcomes of the act of measurement.

Let us consider now some physical system $\mathcal{S}$ and let O_1 and O_2 be two physical observers (for example position and momentum) associated with $\mathcal{S}$. The states of the physical system $\mathcal{S}$ are vectors in a complex vector space $\mathcal{H}$ with the properties of a Hilbert space, whereas the physical observables will be represented by operators denoted by the same symbols which are Hermitian, i.e. $O_i^\dagger = O_i$. Alternatively, any pure state of the physical system is given by a corresponding density matrix ρ. The state at the initial time t_0 is denoted by $\rho_0(t_0)$. We will suppose that the operators O_1 and O_2 are incompatible operators, i.e. they do not commute under the pointwise multiplication of operators, namely $O_1 \cdot O_2 \neq O_2 \cdot O_1$.

Next, we will measure the observable O_1 at the instant t_1 to find the value (eigenvalue) p_1. This measurement is represented on the Hilbert space $\mathcal{H}$ by a Hermitian operator P_1 which is also an idempotent, i.e. $P_1^2 = P_1$. The operator P_1 is a projection operator on the (subspace) eigenspace of the Hilbert space associated with the eigenvalue p_1. The probability of obtaining the eigenvalue p at time t_1 if the state of the system is prepared at the time t_0 to be in the density matrix $\rho_0(t_0)$ is given by Born's rule

$$\mathcal{P}_1 = \text{tr}(P_1(t_1)\rho_0(t_0)). \qquad (2.4)$$

After the first measurement the initial density matrix $\rho_0(t_0)$ collapses to the density matrix $\rho_1(t_1)$ associated with the eigenvalue p_1 given by von Neumann's rule

$$\rho_1(t_1) = \frac{P_1(t_1)\rho(t_0)P_1(t_1)}{\mathrm{tr}(P_1(t_1)\rho(t_0))}. \tag{2.5}$$

Next, we measure the observable O_2 at the instant t_2 to find the eigenvalue p_2. This second measurement is again represented with a projection operator P_2 which projects on the eigenspace of the Hilbert space $\mathcal{H}$ associated with the eigenvalue p_2. The conditional probability of obtaining the second measurement, provided that the first measurement has been performed, is given by Born's rule

$$\mathcal{P}_2 = \mathrm{tr}(P_2(t_2)R_1(t_1)). \tag{2.6}$$

Thus, the probability of obtaining the first measurement $P_1(t_1)$ at time t_1 and then the second measurement $P_2(t_2)$ at time t_2 is given by the product of the conditional probability $\mathcal{P}_2$ and the first probability $\mathcal{P}_1$, namely

$$\mathcal{P} = \mathrm{tr}(P_2(t_2)P_1(t_1)\rho_0(t_0)P_1(t_1)P_2(t_2)). \tag{2.7}$$

Generalization of this result is straightforward. The probability of obtaining the measurements $P_1(t_1)$, $P_2(t_2)$, ... , $P_n(t_n)$ at the successive instants of time $t_1, t_2, \ldots, t_n$, if the state of the system is prepared at the initial instant t_0 to be in the density matrix $\rho_0(t_0)$, is given by the generalized Born's rule

$$\mathcal{P} = \mathrm{tr}(P_n(t_n)\ldots P_2(t_2)P_1(t_1)\rho_0(t_0)P_1(t_1)P_2(t_2)\ldots P_n(t_n)). \tag{2.8}$$

The set of projectors $P_1(t_1)$, $P_2(t_2)$, ... , $P_n(t_n)$ at the instants $t_1, t_2, \ldots, t_n$ defines what we call a quantum history [12].

2.2.2 The quantum Zeno effect and the collapse postulate

The quantum Zeno effect [13] is one of the greatest effects in quantum physics due to its intimate connection to time and consciousness. It asserts the cancellation of motion under continuous quantum measurement and as a consequence it provides an almost direct test of the reduction/collapse postulate.

It is called the Zeno effect because of the intriguing similarity with the Zeno paradoxes of antiquity (which Russell called 'immeasurably subtle and profound' [14]) which are due to Zeno of Elea and his teacher Parmenides. In summary, according to Zeno and Parmenides, there is no motion, time, change, multitude or infinity in the actual world and everything of that nature is illusionary.

Zeno (as recounted by Aristotle in his physics) provided four paradoxes in defense of his teacher's ideas. The arrow paradox for example goes as follows. In order for the arrow in flight to move it must change the position it occupies in space. But at any instant of time the arrow is neither moving to where it is nor it is moving to where it is not. It cannot move to where it is because it is already there and it cannot move to where it is not because there is no elapsed time during the instant of time under consideration. Hence at any instant of time the arrow is not moving, i.e.

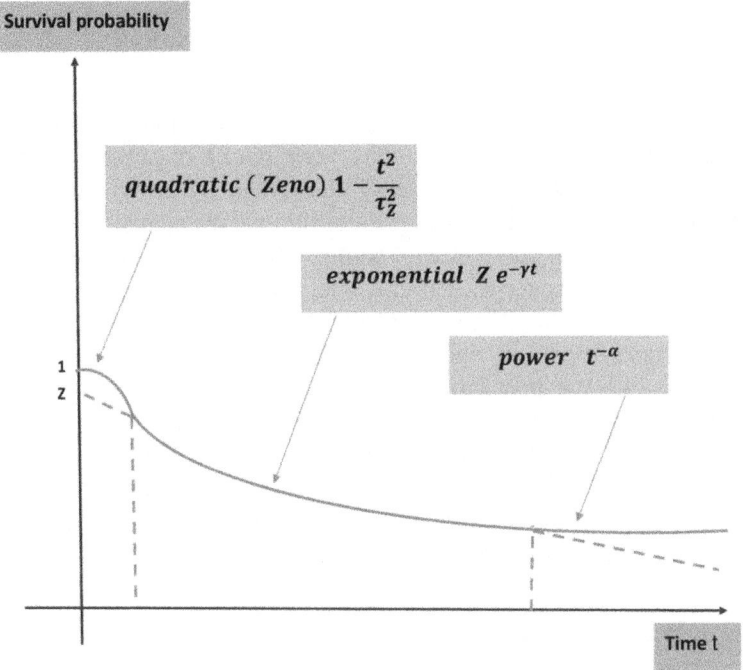

Figure 2.2. The quadratic Zeno region at short times.

motion does not occur, and if motion does not exist then time does not exist. The quantum Zeno effect is effectively saying that there is no unitary evolution in time under a repeated measurement. See the pedagogical presentation in [15].

Sudarshan and Misra in 1977 considered the survival probability which divides into three regimes: (i) a quadratic behavior, (ii) an exponential decay and then (iii) a power law behavior, as shown in figure 2.2.

If the system starts at $t_0 = 0$ from ψ_0 then the probability $p(\delta t)$ of finding the system at a time δt still in ψ_0 is

$$p(\delta t) = 1 - \frac{\delta t^2}{\tau_Z^2}. \tag{2.9}$$

What is the probability of finding the system in its initial state ψ_0 after N continuous measurements?

After the first measurement the unitary evolution of the system starts anew from ψ_0 if the system was found in this state (collapse postulate). The probability that the second measurement will reveal the system to be still in ψ_0 will be given by $p(\delta t)^2$ and not $p(\delta t)$. This second measurement will again collapse the state back onto the initial state if the system was found there and the whole process repeats.

The probability of finding the system in its initial state ψ_0 after N continuous measurements (because of the collapse postulate) is then given by

$$p(t) = p(\delta t)^N = \exp\left(\frac{t^2}{N\tau_Z^2}\right) \longrightarrow 1, \quad N \longrightarrow 1. \tag{2.10}$$

Thus if we perform an increasing number of quantum measurements to check whether or not the system is still in its initial state the chances of actually finding it there become more certain. In some precise sense, the system is frozen in its initial state, i.e. the unitary time evolution is halted by the continuous measurement of its state.

Among the most recent experimental confirmations of the quantum Zeno effect is one by means of a real space measurement of atomic motion, by Patil, Chakram, and Vengalattore [16]. The atoms in an ultracold gas will arrange themselves in a lattice and their velocity is vanishingly small. However, by the Heisenberg uncertainty principle the position and the velocity are conjugate variables. Thus the uncertainty in the position of any given atom is very large and as a consequence the atom can be anywhere in the lattice with equal probability due to quantum tunneling. By subjecting the gas to a continuous measurement (by illuminating them with an imaging laser which causes them to fluoresce) it is observed that quantum tunneling is completely suppressed.

2.3 Entanglement entropy

2.3.1 The reduced density matrix

In quantum mechanics it is shown by the Einstein, Podolsky, Rosen (EPR) experiment, for example, that entanglement is at odds with locality. The action (due to a measurement) seems to propagate with an infinite velocity and although it cannot carry any energy we are left in an uncomfortable position. Entanglement as opposed to energy is not conserved and there are degrees of entanglement. Mathematically, entanglement means that the vector state is not separable, i.e. it cannot be written as a tensor product.

Quantum entanglement is measured by entropy or more precisely by entanglement entropy. However, entropy has actually two sources: statistical and quantum.

1. *The statistical/thermal entropy*: The thermal or Boltzmann entropy of a macroscopic state is the logarithm of the number n of microscopic states consistent with this state. Thus this entropy measures the lack of resolution, i.e. the fact that a large number of microscopic configurations correspond to the same macroscopic thermodynamical state. The thermal entropy is defined in terms of the Boltzmann density matrix $\rho_{\text{ther}} = \exp(-\beta E)/Z$ by

$$\begin{aligned} S_{\text{ther}} &= - \operatorname{tr} \rho_{\text{ther}} \log \rho_{\text{ther}} \\ &= \log n. \end{aligned} \tag{2.11}$$

   The second equality holds if the microstates are equally probable.
2. *Entanglement entropy*:
    - *Measurement*: In quantum mechanics there is another source of entropy associated with the restriction of observers who are performing the experiments, to finite volume. Indeed, a typical observer performing an

experiment on a closed system, which is supposed to be in a pure ground state $|\Psi\rangle$, will only be able to access a particular subsystem, i.e. a partial set of the relevant observables such as those with support in a restricted volume.

We will denote the accessible subsystem by A (where the observers are restricted) and the inaccessible subsystem by B. The total system $\Sigma = A \cup B$ is in a pure ground state $|\Psi\rangle$. See figure 2.3.

- *Reduced density matrix*: The state of the system will be given by a mixed density matrix ρ and the entropy will measure the correlation between the inaccessible subsystem B and the accessible part A of the closed system. The total Hilbert space is $\mathcal{H}_{\text{Tot}} = \mathcal{H}_A \otimes \mathcal{H}_B$.

The observer who cannot access the subsystem B will describe the total system by the reduced density matrix (obtained by tracing over the inaccessible degrees of freedom)

$$\rho_{\text{Red}} \equiv \rho_A = \text{tr}_B \rho_{\text{Tot}}. \tag{2.12}$$

In other words, we trace (integrate) over the inaccessible subsystem B, i.e. we take the average over the inaccessible degrees of freedom.

- *The mixed versus pure states*: The reduced density matrix is an incoherent (mixed) superposition (statistical ensemble, classical probabilities, no interference terms, random relative phases). It is not an idempotent and it satisfies $\text{Tr}\,\rho^2 < 1$.

In contrast, a pure state is a vector in the Hilbert space which is a coherent superposition (interference terms, coherent relative phases) represented by a projector.

Mixed states are relevant if the exact initial state vector is unknown.

- *Entanglement entropy*: The entropy of the subsystem A which measures the correlation between the inaccessible subsystem and the accessible part of the closed system is defined by the von Neumann entropy of this reduced density matrix, namely

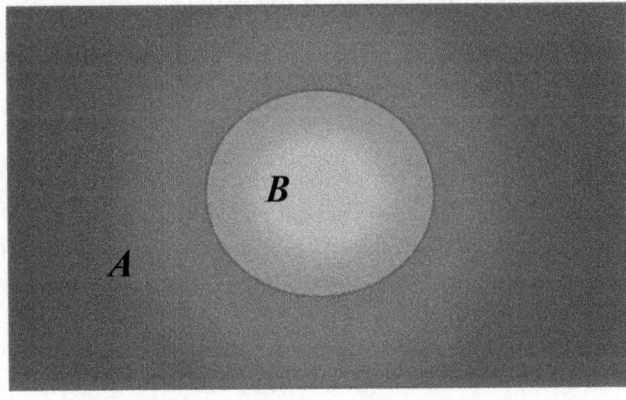

Figure 2.3. The accessible and inaccessible regions.

$$S_{\text{Red}} \equiv S_A = -\text{tr}_A \rho_A \log \rho_A = -\sum_i \rho_i \log \rho_i. \qquad (2.13)$$

Thus, entanglement entropy is the logarithm of the number of microscopic states of the inaccessible subsystem B which are consistent with observations restricted to the accessible subsystem A, together with the assumption that the total system is in a pure state. It measures the degree of entanglement between A and B. This is different from the thermodynamic Boltzmann entropy.

- *Properties*: The entanglement entropy satisfies the following properties. For three subsystems A, B and C which do not intersect each other we have the so-called strong subadditivity relations

$$\begin{aligned} S_{A+B+C} + S_B &\leq S_{A+B} + S_{B+C} \\ S_A + S_C &\leq S_{A+B} + S_{B+C}. \end{aligned} \qquad (2.14)$$

By choosing B empty in the above relations we obtain

$$S_{A+B} \leq S_A + S_B. \qquad (2.15)$$

The mutual information is defined by

$$I(A, B) = S_A + S_B - S_{A+B} \geq 0. \qquad (2.16)$$

If we choose B to be the complement of A then

$$S_A = S_B \Rightarrow S_{A+B} \leq 2S_A. \qquad (2.17)$$

Hence the entanglement entropy is not an extensive quantity.

- *Examples*: For a pure (separable) state, i.e. when all eigenvalues with the exception of one vanish, we obtain $S = 0$. For mixed states we have $S > 0$.

In the case of a totally incoherent mixed density matrix in which all the eigenvalues are equal to $1/N$, where N is the dimension of the Hilbert space, we obtain the maximum value of the von Neumann entropy given by

$$S_{\text{Red}} = S_{\max} = \log N. \qquad (2.18)$$

In the case that ρ is proportional to a projection operator onto a subspace of dimension n we find

$$S_{\text{Red}} = \log n. \qquad (2.19)$$

In other words, the von Neumann entropy measures the number of important states in the statistical ensemble, i.e. those states which have an appreciable probability. This entropy is also a measure of the degree of entanglement between subsystems A and B and hence its other name, entanglement entropy.

- *Information*: The von Neumann entropy $S_V \equiv S_A$ is not additive as opposed to the thermal entropy S_B defined with respect to the Boltzmann distribution. We have $S_B \geq S_V$, i.e. the Boltzmann thermal entropy (coarse grained, macroscopic) is always greater or equal to von Neumann entanglement (fine grained, microscopic) entropy.

The amount of information is the difference

$$I = S_B - S_V. \tag{2.20}$$

If $\Sigma = A$ then there is no entanglement and the amount of information is maximal, i.e. $S_V = 0 \Rightarrow I = S_B$. If $A < <B$ then in this case the amount of information is zero, i.e. $S_V = S_B$. Equivalently, if $A < <B$ then the entanglement entropy becomes maximal equal to the thermal entanglement. Note that the von Neumann entropy of the total system is zero, namely $S_{Tot} = \text{tr} \rho_{Tot} \log \rho_{Tot} = 0$ since there is no inaccessible part here.

2.3.2 Entanglement entropy in quantum mechanics and quantum field theory

For details of the formalism used here we refer the reader to [17]. We will consider a Hamiltonian of the form

$$H = \frac{1}{2} \sum_{A,B} (\delta_{A,B} \pi^A \pi^B + V_{AB} \varphi^A \varphi^B). \tag{2.21}$$

In this equation V is a real symmetric matrix with positive definite eigenvalues. The normalized ground state of this model is given in the Schrödinger representation by

$$\langle \varphi^A | 0 \rangle = \left[\det \frac{W}{\pi} \right]^{1/4} \exp\left[-\frac{1}{2} W_{AB} \varphi^A \varphi^B \right]. \tag{2.22}$$

W is the square root of the matrix V. The corresponding density matrix is

$$\rho(\varphi, \varphi') = \left[\det \frac{W}{\pi} \right]^{1/2} \exp\left[-\frac{1}{2} W_{AB} (\varphi^A \varphi^B + \varphi'^A \varphi'^B) \right]. \tag{2.23}$$

If we suppose that the field degrees of freedom φ^α, $\alpha = \overline{1, n}$, are inaccessible then the correct description of the state of the system will be given by the reduced density matrix in which we integrate out these inaccessible degrees of freedom, namely

$$\rho_{\text{red}}(\varphi^{n+1}, \varphi^{n+2}, \ldots, \varphi'^{n+1}, \varphi'^{n+2}, \ldots) = \int \prod_{\alpha=1}^{n} d\varphi^\alpha \rho(\varphi, \varphi'). \tag{2.24}$$

The entanglement entropy is the associated von Newman entropy of ρ_{red} defined by $S = -\text{Tr} \rho_{\text{red}} \log \rho_{\text{red}}$. The entanglement entropy for any Hamiltonian of the form (2.21) can be shown to be given by [17]

$$S_{\text{ent}} = \sum_i \left[\log\left(\frac{1}{2}\sqrt{\lambda_i}\right) + \sqrt{1 + \lambda_i} \log\left(\frac{1}{\sqrt{\lambda_i}} + \sqrt{1 + \frac{1}{\lambda_i}}\right) \right]. \tag{2.25}$$

The λ_i are the eigenvalues of the following matrix

$$\Lambda_{i,j} = -\sum_{\alpha=1}^{n} W_{i\alpha}^{-1} W_{\alpha j} \qquad (2.26)$$

$W_{\alpha j}$ and $W_{i\alpha}^{-1}$ are elements of W and W^{-1}, respectively, with i, j running from $n + 1$ to $\mathcal{N}$ and α from 1 to n, i.e. Λ is an $(\mathcal{N} - n) \times (\mathcal{N} - n)$ matrix and i, j run from $n + 1$ to N.

The calculation of entanglement entropy in conformal field theory is more involved but is based on the same formula $S = -\mathrm{Tr}\,\rho_{\text{red}} \log \rho_{\text{red}}$. See [18] and references therein.

In a QFT on a $(d + 1)$-dimensional manifold $\mathbf{R} \times \mathbf{N}$, where $d \geqslant 2$ and $\mathbf{N} = A \cup B$, it is found that entanglement entropy:
1. depends only on the geometry of A (this is why entanglement entropy is also called geometric entropy).
2. is UV divergent and hence the continuum theory should be regularized by a lattice a.
3. is proportional to the area of the boundary ∂A of A since the entanglement between A and B occurs strongly and obviously on the boundary.

We have explicitly [17, 19]

$$S_A = \gamma \cdot \frac{\mathrm{Area}(\partial A)}{a^{d-1}} + \text{subleading terms}. \qquad (2.27)$$

This entanglement entropy formula (which includes UV divergences and is proportional to the number of matter fields) is very similar to the Bekenstein–Hawking formula (which does not include UV divergences and is not proportional to the number of matter fields). In fact the quantum corrections to the Bekenstein–Hawking black hole entropy in the presence of matter fields is given by entanglement entropy [20–23].

2.4 EPR and Bell's theorem
2.4.1 The Einstein, Podolsky, Rosen experiment
The celebrated Einstein, Podolsky, Rosen (EPR) gedanken experiment [24] is based on two assumptions:
- *EPR1 or classical realism*: In other words, the world, or more precisely its properties, really exist independently of any measurement. Thus, a physical quantity is real if its value can be predicted with certainty (hence the need for hidden variables since the Schrödinger equation does not permit this) without disturbing the system being measured (there should be no entanglement, which Einstein termed 'spooky action at a distance').
- *EPR2 or locality*: The physical properties of a system A should be independent from the physical properties of a spatially separated system B (no entanglement again). This assumption is related closely to relativity in an almost obvious sense!

These two assumptions led them directly to the conclusion that the quantum wave function given by the solution of the Schrödinger equation is an incomplete description of physical reality and thus hidden variables are needed.

Bohr, the father of the orthodox and the Copenhagen interpretations, was opposed to EPR1 (realism) more than to EPR2 (locality). Bell then showed that the two EPR assumptions lead directly to what we now call Bell's inequality, which is violated badly by quantum mechanics [25] and nature (the famous Aspect experiment [26]).

Putting it differently, one of the two EPR assumptions or both are at odds with quantum mechanical predictions. Bell himself rejected EPR2, i.e. locality or more precisely local causality, which is also the view of the majority of physicists and philosophers with the exception perhaps of the supporters of the consistent (decoherence) histories approach who reject classical realism in favor of so-called quantum realism (the single framework rule) [12].

Thus the world according to the views of the majority of physicists and philosophers who understand quantum mechanics in this particular way is certainly not local. It may even not be classically real and there is even an implicit danger to free will and/or causality.

2.4.2 Theorems of quantum philosophy

The three fundamental theorems of quantum philosophy are:

1. *The Kochen–Specker theorem (1967)*: The Kochen–Specker theorem [27] states simply that no hidden variable contextual description of quantum mechanics is possible.

    This theorem depends on the no-contextuality requirement: the results of a given measurement which are predicted by the underlying state (wave function and hidden variables) do not depend on what other measurements are being performed on the system. In the contextual hidden variable theories (such as Bohm's interpretation [28]) the result of a given measurement depends on the state and on the other measurements being performed on the system.

2. *Bell's theorem (1965)*: This states that hidden variable theories can only be non-local [25, 29]. This is the most fundamental of all these theorems.

3. *The Greenberger–Horne–Zeilinger (GHZ) theorem (1989)*: The GHZ theorem [30] is a generalization of Bell's theorem which is mid-way between the algebraic no hidden variable theorem (combinatorial considerations) of Kochen and Specker and the statistical hidden variable theorem (multi-particle considerations) of Bell. This situation is termed Bell without statistics by [31].

    The GHZ theorem involves a maximally entangled tripartite system as opposed to the maximally entangled bipartite system considered in Bell's theorem. As in Bell's theorem, the GHZ theorem rules out local hidden variable theories. Both Bell and GHZ rely on the absence of advanced action.

These three major theorems are the most difficult objections to the ignorance interpretations of quantum mechanics which assume that quantum mechanics is incomplete and thus it should be supplemented by hidden variables. These theorems show that any hidden variable description of quantum mechanics must be both contextual and non-local. Only non-local and contextual hidden variable interpretations such as Bohm's interpretation [28] can escape these no-go theorems.

A fourth theorem worth mentioning is the free will theorem of Conway and Kochen [32, 33]. This is a mixture of the Kochen–Specker and Bell's theorems and leads to some puzzling consequences for the problem of free will (see also [34, 35]).

2.4.3 Bell's theorem

The state vector (or wave function) of a physical system only permits us to calculate the probabilities of various possible outcomes in a given measurement. This is the orthodox position. The question one can then ask immediately: did the physical system have all along, i.e. before the act of measurement, the value found after measurement?

Einstein, and all those who believe in local realism, would answer this question in the affirmative. This is the substance of the famous EPR paradox [24] (see also the nice discussion in [36]). Thus, in this view, the system has the measured value long before the act of measurement had took place and quantum mechanics is simply an incomplete theory, as it can only allow us to calculate probabilities. In other words, there must exist extra hidden variables which, together with the wave function, will allow us to predict precisely the behavior of the physical system exactly and deterministically before the measurement.

This picture assumes therefore implicitly/explicitly: (i) reality, (ii) locality and (iii) free choice. Bell's theorem [29] shows that these extremely reasonable assumptions are not compatible. It states precisely that no physical theory based on local hidden variables can reproduce all the predictions of quantum mechanics [29, 37]. Bell's theorem is justifiably the most profound result in physics and some may even go further and consider it 'the most profound discovery of science' [38].

Most interpretations of quantum mechanics will relax either the assumption of reality, or the assumption of locality, but rarely the assumption of free choice (Bell himself discussed super-determinism while Price [31] considered backward causation in connection with the transactional interpretation of quantum mechanics [39]).

The majority (I guess) view (the Bohr and the Copenhagen school [7]) states, on the other hand, that the state of the system actually does not exist before measurement (see also [40]), which is what seems to be confirmed by Aspect's experiment [26].

We consider the set-up of the EPR thought experiment of 1937 as re-imagined by Bohm [41] and Bell [37], which is given by a neutral pion particle decaying into an electron and positron pair. This is given by the decay process

$$\pi_0 \longrightarrow e^- + e^+. \qquad (2.28)$$

The electron and the positron fly away in opposite directions due to the conservation of momentum, i.e. since the pion decays at rest. The state vector of the system electron + positron is a maximally entangled spin state (known as Bell's state) given by the singlet state

$$\frac{1}{\sqrt{2}}(|+\rangle|-\rangle - |-\rangle|+\rangle). \tag{2.29}$$

Thus, if the spin of the electron is up, then the spin of the positron must be down, and if the spin of the electron is down, then the spin of the electron must be up. The probability for each possibility is given by 1/2.

We now leave the electron and positron to fly away from each other, in this entangled state, until their distance apart becomes as large as desired, perhaps of the order of the diameter of the observable Universe, i.e. we allow the two systems to become causally disconnected. Then, we (or rather Alice) perform a spin measurement on, say, the electron.

The idea behind this gedanken experiment is that the electron and positron are allowed to separate as far apart as possible, that there is strictly no possible causal mutual influence between the two, and hence when Alice performs our measurement on the electron on this side of the Universe, the measurement on the spin of the positron (by Bob) on the other side of the Universe can be supposed to be a completely uncorrelated measurement, yet because of entanglement and collapse this measurement is really completely determined.

What does quantum mechanics say precisely about such a situation? The measurement of the spin of the electron by Alice will yield the values with equal probability 1/2, and similarly the measurement of the spin of the positron by Bob will yield the values with equal probability 1/2.

However, because the system is found in an entangled state, the measurement of the spin of the electron by Alice will collapse the wave function, and as a consequence the spin of the positron can be determined with certainty, regardless of the measurement of Bob. For example, if when we measure the spin of the electron we find spin up, the wave function collapses which means the spin of the positron, which is on the other side of the Universe, must be down with certainty. And if we find the spin of the electron to be down, then we know that the spin of the positron must be up without any further measurement. Thus, the effect of the collapse propagates instantaneously even from one side of the Universe to the other side, and this is what Einstein has called 'spooky action at a distance', which goes against the spirit of relativity.

The solution according to Einstein lies in the reality of the wave function, i.e. the spin of the electron is well-defined even before measurement. In other words, it is really either up or down before measurement. And thus quantum mechanics in allowing us to only calculate probabilities is simply an incomplete theory. Hence the need for hidden variables, i.e. the wave function ψ must be supplemented by an extra variable (hidden variable) λ which allows a complete specification of the state of the system.

As discussed above, Bell has shown that any local deterministic hidden variable theory cannot reproduce all the predictions of quantum mechanics. This goes as follows (we follow the simplified presentation of [42]). We start by following Bell in measuring the spin S_{1a} of the electron in an arbitrary direction $\vec{a}$, while measuring the spin S_{2b} of the positron in another arbitrary dimension $\vec{b}$. See figure 2.5.

According to the rules of standard quantum mechanics outlined above, the expectation value of the product of the two spins, namely $\langle S_{1a} S_{2b} \rangle$, is given by the scalar product

$$P(a, b) = -\vec{a} \cdot \vec{b}. \tag{2.30}$$

But according to Einstein, and all those who have a natural inclination toward local realism and hidden variables, the wave function ψ comes with a hidden variable λ given by some probability density $\rho(\lambda)$ satisfying as usual

$$\rho(\lambda) > 0, \quad \int d\lambda \rho(\lambda) = 1. \tag{2.31}$$

This is the assumption of realism.

We will further assume locality, which here means the requirement that the directions $\vec{a}$ and $\vec{b}$ are freely and independently chosen, and which also means in general that physical actions cannot propagate faster than the speed of light, and thus measurements made at places which are space-like separated cannot influence each other.

Thus we will also assume that the measurements S_{1a} and S_{2b} of the spins of the electron and positron are given by two functions f and g which can only take the two values ± 1, i.e. $f(\vec{a}, \lambda) = \pm 1$ and $g(\vec{b}, \lambda) = \pm 1$, such that when the two spins are aligned we obtain precisely anti-correlated measurements, namely

$$f(\vec{a}, \lambda) = -g(\vec{a}, \lambda). \tag{2.32}$$

The expectation value $P(a, b)$ for the product of two spins should then be given by the equation

$$P(a, b) = \int \rho(\lambda) f(\vec{a}, \lambda) g(\vec{b}, \lambda) d\lambda. \tag{2.33}$$

Any deterministic local hidden variable theory with these general properties will then give an expectation value $P(a, b)$ satisfying, for three arbitrary directions $\vec{a}, \vec{b}, \vec{c}$, the inequality

$$|P(a, b) - P(a, c)| < 1 + P(b, c). \tag{2.34}$$

This very simple result is the celebrated Bell's inequality.

As it turns out, this hidden variable's result is quite incompatible with the above quantum mechanical prediction, i.e. with $P(a, b) = -\vec{a} \cdot \vec{b}$. For example, if $\vec{a}$ is perpendicular to $\vec{b}$, and $\vec{c}$ makes a 45 degree angle with $\vec{a}$ and $\vec{b}$, we obtain [42]

$$P(a, b) = 0, \, P(a, c) = P(b, c) = -0.7. \tag{2.35}$$

This is clearly not satisfied by Bell's inequality.

To highlight the severity of this violation we consider a simple problem from set theory and logic. Let A, B and C be three properties with corresponding sets indicated by the Venn diagrams in figure 2.4.

Let $\mathcal{N}_1$ be the number of objects which have property A but not property B, i.e. $\mathcal{N}_1 = N_1 + N_2$. And let $\mathcal{N}_2$ be the number of objects which have property B but not

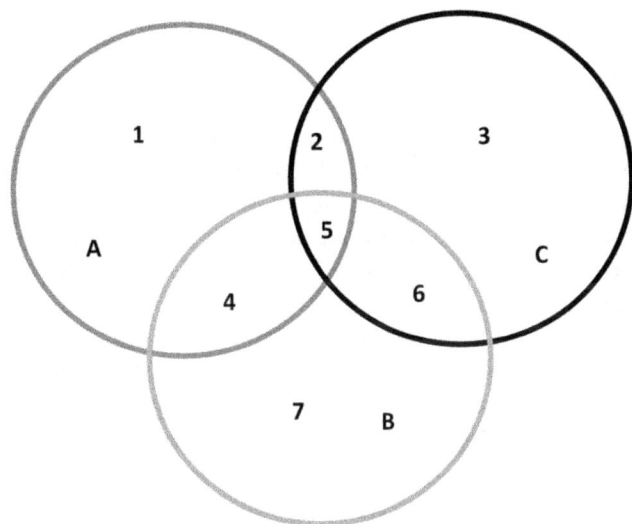

Figure 2.4. Bell's inequality with Venn sets.

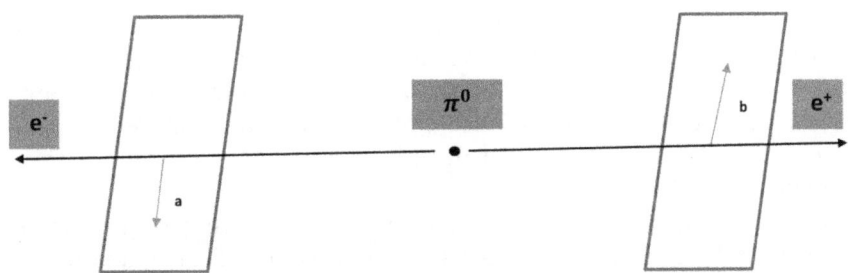

Figure 2.5. The EPRBB experiment.

property C, i.e. $\mathcal{N}_2 = \mathcal{N}_7 + \mathcal{N}_4$. And let $\mathcal{N}_3$ be the number of objects which have property A but not property C, i.e. $\mathcal{N}_3 = \mathcal{N}_1 + \mathcal{N}_4$.

It is then trivial to show (the proof is visual from the Venn diagrams in figure 2.4) that

$$\mathcal{N}_1 + \mathcal{N}_2 \geqslant \mathcal{N}_3. \tag{2.36}$$

This is Bell's inequality in this context. We are saying that this logical inequality is violated badly by quantum mechanics and nature.

Thus Bell has shown in a very simple way that if Einstein's local realism is correct, then quantum mechanics is not merely incomplete but it is plainly wrong. On the other hand, if quantum mechanics is correct, then no local hidden variable theory can be made consistent with quantum mechanics. The Aspect, Grangier and Roger experiment of 1982 has shown decisively that Bell's inequalities are violated in reality and quantum mechanics predictions are fully vindicated.

For some recent work on the violations of Bell's inequality see [43]. Thus nature at the most fundamental of levels seems to not really be real and is perhaps also non-local (figures 2.4 and 2.5).

2.5 Decoherence and the measurement problem

The process of quantum measurement is one of the most fundamental aspects of quantum theory, which involves in an essential way many profound quantum effects such as the collapse of the wave function, quantum entanglement and decoherence. This makes it a very hard (in fact virtually intractable) problem (so far).

The phenomena of decoherence [44–46] in particular involves the unavoidable interaction between the quantum system and the environment, which looks very much like a measurement, and thus the state of the quantum system becomes entangled with the state of the environment, and this in turn causes what looks like a collapse of the wave function, i.e. the reduction of a quantum pure state to a statistical mixture causing a loss of coherence. In other words, decoherence explains (I think very well) how classicality emerges from the underlying quantum nature. What remains debatable, and for some a dubious assertion, is the claim that the quantum measurement is nothing more and nothing less than the decoherence due to the coupling of the quantum system to the environment (including the measurement devices, brains and minds).

Thus the physics of decoherence is not controversial, but the claim that the quantum measurement is simply decoherence is still very much open to debate. For one, the Copenhagen school still maintains that the state of the system does not exist before measurement, while the many-worlds interpretation maintains that it exists in various branches of the many-worlds, and the consistent histories approach maintains that the state is simply indeterminate. It seems that decoherence does not favor any of these positions over the others.

In summary, we have:

- Decoherence is the unavoidable interaction between the quantum system (open system, i.e. the Schrödinger equation is inapplicable) and its environment. Coherence of the wave function is destroyed and quantum entangled pure states are turned into classical statistical mixtures.
- The boundary between the quantum (linear combination) and the classical (determinism) is therefore dynamically determined by decoherence and not by the act of measurement. This in fact is an interpretation of the collapse of the wave function!
- The measurement yields an entangled state of the quantum system and the measuring device (detector). This entangled state obeys the Schrödinger equation and is a correlated and non-separable state (Aspect experiment) which violates the Bell inequalities.
- Conclusion: the states of the quantum system in the entangled state do not, and in fact cannot, exist before measurement.

However, an observer who did not inspect the detector will describe the system by a density matrix. The density matrix ρ_e associated with the entangled state is pure (interference terms or coherences). Measurement will take this pure density matrix to a reduced density matrix ρ_r which is mixed (non-unitary, collapse, no coherences).

To illustrate this fundamental point we consider as a system S a spin one-half particle with two states $|+\rangle$ and $|-\rangle$. The detector D clicks if it measures spin up and does nothing otherwise.

The interaction between the system S and the detector D then produces an entangled pure state as follows:

$$|\Phi_i\rangle = (\alpha|+\rangle + \beta|-\rangle)|D-\rangle \Rightarrow \alpha|+\rangle|D+\rangle + \beta|-\rangle|D-\rangle = |\Phi_e\rangle. \qquad (2.37)$$

This pure state defines a pre-measurement or an incompleted measurement and it is described alternatively by the pure density matrix

$$\begin{aligned}\rho_e &= |\Phi_e\rangle\langle\Phi_e| \\ &= |\alpha|^2|+\rangle|D+\rangle\langle+|\langle D+| + |\beta|^2|-\rangle|D-\rangle\langle-|\langle D-| \\ &+ \alpha\beta^*|+\rangle|D+\rangle\langle-|\langle D-| + \beta\alpha^*|-\rangle|D-\rangle\langle+|\langle D+|.\end{aligned} \qquad (2.38)$$

The completed measurement is described by the mixed density matrix (off-diagonal/interference terms are canceled)

$$\rho_r = |\alpha|^2|+\rangle|D+\rangle\langle+|\langle D+| + |\beta|^2|-\rangle|D-\rangle\langle-|\langle D-|. \qquad (2.39)$$

The (discontinuous, irreversible, instantaneous, non-deterministic and non-unitary) transition $\rho_e \rightarrow \rho_r$ is the collapse postulate in the Copenhagen interpretation. But decoherence claims that it is a dynamical process obtained by taking into account the environment.

Yet, there is no known process which effectuates the transition $\rho_e \rightarrow \rho_r$. This is the measurement problem.

We should then consider the system consisting of the quantum system S, the detector D and the environment E. The interaction of the environment with $S + D$ is also described by an entangled state and a pure density matrix as follows:

$$\begin{aligned}|\Phi_e\rangle|E_0\rangle &= (\alpha|+\rangle|D+\rangle + \beta|-\rangle|D-\rangle)\\ |E_0\rangle &\Rightarrow \alpha|+\rangle|D+\rangle|E+\rangle + \beta|-\rangle|D-\rangle|E-\rangle = |\Psi\rangle.\end{aligned} \qquad (2.40)$$

The density matrix of the system $S + D$ is then obtained by tracing over the degrees of freedom in the environment E (which are inaccessible) to obtain the reduced density matrix, namely

$$\begin{aligned}\rho_{S+D} &= \text{Tr}_E |\Psi\rangle\langle\Psi| \\ &= |\alpha|^2|+\rangle|D+\rangle|+\rangle\langle D+| + |\beta|^2|-\rangle|D-\rangle|-\rangle\langle D-| \\ &= \rho_r.\end{aligned} \qquad (2.41)$$

This is the claim of decoherence, or more precisely the claim of those who use decoherence to interpret the collapse or reduction of the state vector.

2.6 The many-worlds formalism

2.6.1 The many-worlds formalism and coherent branching

The many-worlds formalism was introduced in 1957 by Hugh Everett III in his doctoral dissertation under Wheeler [47, 48]. He actually dubbed it the 'relative state formulation' and it is reported that he was in fact dismissive of the term 'many-worlds' introduced by DeWitt and Graham [49] when they revived this formalism in the 1970s.

The only postulate of the many-worlds formalism is the unitary evolution in time given by the Schrödinger equation which is the only admitted process (as opposed to Copenhagen's two processes). The wave function is of a real ontology, not of a merely descriptive value, and the collapse of the wave function never occurs.

In the measurement process the collapse of the wave function is replaced by the splitting or branching (which is a fully reversible and unitary process) of the world which is a literal and direct reflection of the linear superposition principle. The frequency of branching is given precisely by Born's rule.

The many-world formalism is complementary to the Copenhagen interpretation, not contradictory (this is this author's view). This is much stronger than the view (of Susskind) that the Copenhagen interpretation is a very good approximation of the many-worlds formalism. In fact the Copenhagen is thought of as an exact statement with respect to the conscious/zombie observer whereas the many-worlds formalism is an exact statement with respect to a super-observer who does not interfere with the world.

An analogy due to Tegmark [50] is to think of the many-worlds formalism as playing the role of the manifold structure of spacetime in general relativity, whereas the Copenhagen interpretation plays the role of the local flatness observed by every observer around each point in spacetime.

Thus the non-unitary observer-participancy (as Wheeler puts it [4, 51]) or consciousness-causes-collapse (von Neumann–Wigner interpretation [52, 53]) found in the single-world of the Copenhagen interpretation is replaced by a unitary many-worlds formalism (or a many-minds formalism [54, 55] in which the unitary branching occurs in the mind and not in the world). In some sense the extreme view of a non-unitary efficacious role of consciousness in quantum mechanics (quantum dualism in a single-world [56]) is dual/complementary to a no less extreme view of a unitary reality with many coherent and parallel worlds (physicalism in a many-worlds scenario).

2.6.2 Schrödinger's cat and quantum immortality

Schrödinger's cat experiment is definitely among the greatest quantum experiments ever devised. It was introduced by Schrödinger in 1935 to highlight the conceptual problems with the Copenhagen interpretation [57, 58]. The physical system here is a conscious cat. Thus there is an object (the cat), a subject (the observer, i.e. mind or the detector), and the inaccessible and unavoidable environment.

The object and the subject are related through the quantum measurement. If no measurement is made on the cat then the state of the cat is a linear superposition given by

$$|\text{cat}\rangle = \frac{1}{\sqrt{2}}(|\text{alive}\rangle + |\text{dead}\rangle). \tag{2.42}$$

- Question 1: Is the cat dead and alive at the same time or is she neither dead nor alive?
- Question 2: When a measurement is performed what do we find?

The many-worlds answer (no collapse, branching, wave function has an objective reality) is given by the state vector

$$\frac{1}{\sqrt{2}}(|\text{alive}\rangle|\text{happy}\rangle + |\text{dead}\rangle|\text{dead}\rangle). \tag{2.43}$$

Thus there is a branch of the many-worlds in which the cat is alive and another branch in which the cat is dead and the two branches are coherent. It is decoherence that destroys this linear superposition between the two branches and turns them into parallel (independent) worlds. Thus decoherence is precisely the relation between object and environment on one hand and subject and environment on the other hand. In some sense decoherence acts as a measurement.

So the cat is alive in one world and is dead in another world and the two worlds are both genuinely real. This is to be contrasted with the Copenhagen interpretation, which states that these states do not exist before measurement.

In summary, the Copenhagen interpretation destroys the objectivity of classical reality by giving the subject a special role (but reality is really not classical but quantum and the quantum dualism describes it perfectly as such!).

On the other hand, the many-worlds formalism maintains the objectivity of the classical reality which is formed (we have to accept it) of an infinite number of coherent branches and decohered parallel worlds. No special role is given to the subject. In some sense the many-worlds formalism is an external view in which the mathematics has an objective reality whereas the Copenhagen is an internal view in which the mathematics is a representation or approximation of reality [50].

Another related gedanken experiment which is profoundly puzzling is quantum immortality, proposed by Tegmark [50]. It is claimed to be the only experiment which can discriminate between the Copenhagen and many-worlds approaches.

In the current case Schrödinger's cat is replaced by Schrödinger's experimenter. A quantum gun is prepared in the state

$$\frac{1}{\sqrt{2}}(|\text{up}\rangle + |\text{down}\rangle). \tag{2.44}$$

The trigger will be pulled and the gun fires if the measurement of a qubit yields the value -1, otherwise nothing happens. We repeat n times. According to the Copenhagen interpretation the probability of survival after n steps is obviously $1/2^n$.

The state after the first measurement according to the many-worlds formalism is

$$\frac{1}{\sqrt{2}}(|\text{up}\rangle|\text{alive}\rangle + |\text{down}\rangle|\text{dead}\rangle). \tag{2.45}$$

We suppose (i) oblivion (no physical consciousness after death) and (ii) continuity of identity (time between measurements is much smaller than the time of human consciousness).

We immediately conclude that the experimenter will find herself alive at each step, i.e. probability of survival is 1. This is because there is one conscious person after and before the experimenter, her identity is continuous, and the other persons in the other branches have all suffered oblivion.

In fact, the experimenter in most branches is dead but there exists one branch where the experimenter survives and, because the assumptions of oblivion and continuity of identity hold, the experimenter never dies and she acts as if she is immortal. However, this experimenter will be the only person who knows this and thus she can objectively discriminate between the Copenhagen interpretation and many-worlds formalism, favoring the many-worlds formalism.

2.6.3 The many-minds interpretation

The many-minds interpretation is a very close relative of the many-worlds interpretation which involves the following crucial modification/twist: the split or branch of the world into parallel branches when a quantum measurement is performed is shifted to a split or branch of the mind into parallel minds. In both interpretations, it is assumed that quantum mechanics as it stands is a complete theory of nature and that there is no collapse of the wave function under measurement, which is what is accounted for by the splitting into branches.

This means, in particular, that the fundamental law is given by the Schrödinger equation alone, and the relationship between branching and relative frequencies, which is ill defined *a priori* in both pictures, should be given for consistency by Born's rule.

The most important versions of the many-mind interpretation are as follows:
- *The Albert and Loewer theory* [54]: This is an in intrinsically dualistic theory which assumes in the words of Albert 'that every sentient physical system is associated not with a single mind but rather with a continuous infinity of minds'. However, in this theory, there is no supervenience of brane states and mind states.
- *Lockwood theory* [55]: In this theory there is a complete supervenience of the physical and mental.

In the following we will follow [59].

Before we begin, we mention a few other implications of the Albert–Loewer theory, which is the most important one for us here because of its dualistic character. First, an epistemological implication of the Albert–Loewer theory is the observation that our current experience could be fully compatible with the fact that the Universe has always been in a vacuum state. This seem to me to also be an ontological implication. Another implication of the Albert–Loewer theory is that quantum non-locality is removed completely from the physical and delegated to the mental world. In fact all many-worlds and many-minds interpretations are no-collapse models and they avoid non-locality by claiming that Bell correlations (predicted by Bell's theorem) are not fully objective correlations but they are observer-dependent. Price's critique of these claims reaches the conclusion that these no-collapse models do not really eliminate non-locality but they simply explain it.

The system S under consideration is assumed to be composed of a single electron. We are interested in the measurement of the z component of the spin. The measurement apparatus is denoted by M and the observer is denoted by O. The total system $S + M + O$ is initially prepared in the state

$$|\Psi_0\rangle = (\alpha|-\rangle + \beta|+\rangle) \otimes |\psi_0\rangle \otimes |\phi_0\rangle. \tag{2.46}$$

The state $|\psi_0\rangle$ is the initial state of the apparatus and $|\phi_0\rangle$ is the initial state of the observer's brain. The complete state $|\Psi_0\rangle$ is supposed to obey only the Schrödinger equation. In other words, we assume that there is no collapse.

The measurement interaction between the system S and the measurement apparatus M creates a one-to-one correlation between the states of up and down spins of the electron and the pointer states $|\psi_\pm\rangle$ of the apparatus. Hence the system S and the measurement apparatus M become entangled, i.e. the measurement interactions result in taking the above state $|\Psi_0\rangle$ to the combination

$$|\Psi_1\rangle = (\alpha|-\rangle \otimes |\psi_+\rangle + \beta|+\rangle \otimes |\psi_-\rangle) \otimes |\phi_0\rangle. \tag{2.47}$$

Next we assume that the brain states corresponding to all possible outcomes of all possible experiments form a preferred basis in the brain's Hilbert space. These states correspond to those mental states associated with the conscious perception of the outcomes of the experiments. Let us denote the two brain states associated with the conscious perception of the states of up and down spins of the electron by $|\phi_\pm\rangle$. Then the interaction between the measurement apparatus M and the observer O will take the state $|\Psi_1\rangle$ to the final state

$$|\Psi_f\rangle = \alpha|-\rangle \otimes |\psi_+\rangle \otimes |\phi_+\rangle + \beta|+\rangle \otimes |\psi_-\rangle \otimes |\phi_-\rangle. \tag{2.48}$$

The measurement has no definite result and thus this theory (called the bare theory by Albert) is not complete and it should then be supplemented by extra ingredients.

The many-mind interpretation of Albert and Loewer is a no-collapse interpretation in which we suppose that the bare theory is complete with respect to the physics including the brain. However, regarding the relationship of the brain states $|\phi_\pm\rangle$ to the mental states of the observer O we also assume the following two postulates:
- Each brain state $|\phi\rangle$ is associated at all times with an infinity of non-physical minds.
- The minds do not obey the Schrödinger equation but evolve in time in a stochastic way with a probability given by the Born rule.

Thus we start with an infinity of minds associated with the initial brain state $|\phi_0\rangle$. In some sense the minds are degenerate, all described by the single brain state $|\phi_0\rangle$. Each mind then evolves in a stochastic way to either the state $|\phi_+\rangle$ with the Born probability $|\alpha|^2$ or to the state $|\phi_-\rangle$ with the Born probability $|\beta|^2$. Thus the state $|\Psi_f\rangle$ should be replaced by

$$|\Psi_f(m, n)\rangle = \alpha|-\rangle \otimes |\psi_+\rangle \otimes |\phi_+(m)\rangle + \beta|+\rangle \otimes |\psi_-\rangle \otimes |\phi_-(n)\rangle. \tag{2.49}$$

The notation $|\phi(m)\rangle$ means that the brain state $|\phi\rangle$ corresponds to and is indexed by the subset m of the set of minds. In other words, the description of the post-measurement state includes the quantum states of the system S and of the apparatus M and the subsets of the set of minds in the + and − branches of the superposition.

This interpretation is truly probabilistic since before the divergence of the minds into the branches of the state $|\Psi_f\rangle$ it is fully random which branch each mind will actually follow. The probability is given by the quantum mechanical Born rule. The requirement of an infinite number of minds is put forward in order to avoid (i) the so-called mindless hulk problem and also in order to avoid (ii) Bell's non-locality.

More important is the fact that this interpretation is dualistic in the sense that only subsets of the set of minds (and not individual minds) supervene on the brain states. Thus any m-mind can be exchanged with any n-mind leaving the physics invariant.

The other issue concerns the relationship between branching and relative probability, which is a major problem in the many-worlds interpretation as well. This is solved by simply assuming the Born rule, as shown originally by Everett in 1957. It can then be proved that the probability of each branch on a given tree is given by the quantum mechanical Born rule and that each individual mind performs a classical random walk on this tree with this probability. This does not mean that there exists a non-contextual classical probability distribution which can assign the correct probability to all branches at once in accordance with the violation of Bell's inequality.

Indeed, the probability of the branching must be conditional on the measurement performed. If the probability were pre-determined then the minds will act as hidden variables and they will necessarily violate Bell's inequality, i.e. we have a non-local hidden variables theory. Hence in order to avoid this non-locality we must assume a random distribution of the minds, which is conditional on the given measurement.

In the Lockwood interpretation there is a complete supervenience of the (continuous infinity of) minds on the brain states. Also the minds are supposed to be not stochastic. Thus the final post-measurement state is given by $|\Psi_f\rangle$ and not $|\Psi_f(m, n)\rangle$. In other words, subsets of minds are indexed by brain states as opposed to brain states being indexed by subsets of minds. Thus the fraction of minds in the branch + is proportional to $|\alpha|^2$ whereas the fraction of minds in the branch − is proportional to $|\beta|^2$. On the other hand, the dynamics of minds, which is not random in this case, is not clear and some possibilities are discussed, for example, in [59].

2.7 Bohmian mechanics

2.7.1 A deterministic non-local theory

Bohmian mechanics is the only deterministic, and thus causal, hidden variable interpretation theory of quantum mechanics, which was conceived originally by de Broglie [60] and then really constructed by Bohm [28]. It is the only explicit non-local formulation of quantum mechanics through the introduction of the so-called quantum potential, which thus attempts to reflect, consciously or otherwise, the non-local, i.e. action at a distance, character of physical reality as described by quantum mechanics.

Being non-local means, in particular, that Bohm evades the constraints imposed by Bell's theorem, which was confirmed experimentally, for example, by Aspect *et al*, by considering a non-local hidden variable extension of quantum mechanics.

The state of the system in this interpretation is given by the usual wave function ψ in the Hilbert space, together with the usual generalized coordinates q_i of classical mechanics. As usual, the set of generalized coordinates q_i defines a point $\vec{q}$ called a configuration in configuration space. It can be argued that the wave function is in fact the hidden variable here since it is not measurable as opposed to the measurable positions q_i [61]. A more serious discrepancy is the fact that the q_i are actually the classical positions not the actual quantum positions $\langle \hat{x}_i \rangle$. This discrepancy can be alleviated somewhat if we keep in mind this difference and translate back to the actual quantum positions whenever it is needed.

The evolution of the positions q_i in time is given in terms of the wave function ψ itself by means of the so-called guiding equation. Hence Bohmian mechanics contains, in addition to the usual Schrödinger equation, which governs the evolution of the wave function ψ, the guiding equation which governs the evolution of the configuration $\vec{q}$ in terms of the wave function. The evolution of positions q_i is then in a clear sense guided by the wave function. This is clearly a deterministic system. Thus according to Bohm quantum mechanics is as deterministic as classical mechanics.

As pointed out originally by Bohm himself the predictions of Bohmian mechanics and quantum mechanics should fully coincide. He says in [28]: 'as long as the present general form of Schroedinger's equation is retained the physical results obtained with this suggested alternative interpretation are precisely the same as those obtained with the usual interpretation'.

This is true almost by construction, as we will see. The only possible source of confusion is the existing difference between the generalized coordinates of the system q_i and the quantum positions $\langle \hat{x}_i \rangle$.

We start then with the wave function $\psi = \psi(t, \vec{x})$, which obeys as usual the Schrödinger equation

$$i\hbar \frac{\partial \psi}{\partial t} = H\psi, \quad H = -\frac{\hbar^2}{2m}\vec{\nabla}^2 + V. \tag{2.50}$$

Following Bohm's original derivation we polar decompose the wave function as

$$\psi = R \exp(iS/\hbar). \tag{2.51}$$

We compute immediately

$$i\hbar \frac{\partial \psi}{\partial t} = (i\hbar \frac{\partial \ln R}{\partial t} - \frac{\partial S}{\partial t})\psi \tag{2.52}$$

and

$$H\psi = -\frac{\hbar^2}{2m}\left(\frac{1}{R}\vec{\nabla}^2 R + \frac{2i}{\hbar}\vec{\nabla}S\vec{\nabla}\ln R + \frac{i}{\hbar}\vec{\nabla}^2 S - \frac{1}{\hbar^2}(\vec{\nabla}S)^2 \right)\psi + V\psi. \tag{2.53}$$

By equating the two terms we obtain

$$\frac{\partial S}{\partial t} + \frac{1}{2m}(\vec{\nabla}S)^2 + V - \frac{\hbar^2}{2m}\frac{\vec{\nabla}^2 R}{R} = i\hbar\left(\frac{\partial \ln R}{\partial t} + \frac{\vec{\nabla}S}{m}\vec{\nabla}\ln R + \frac{1}{2m}\vec{\nabla}^2 S\right). \quad (2.54)$$

Now we introduce the velocity operator $\vec{\hat{v}}$ acting on the Hilbert space $\mathcal{H}$ through the correspondence principle, namely

$$\vec{\hat{v}}\psi = \frac{\vec{\hat{p}}}{m}\psi = \frac{1}{m}\frac{\hbar}{i}\vec{\nabla}\psi = \frac{1}{m}\left[\frac{\hbar}{i}\vec{\nabla}\ln R + \vec{\nabla}S\right]\psi. \quad (2.55)$$

In the classical limit $\hbar \to 0$ the action of this velocity operator becomes simply

$$\vec{\hat{v}}\psi = \frac{\vec{\hat{p}}}{m}\psi = \frac{1}{m}\vec{\nabla}S\psi. \quad (2.56)$$

Thus S is Hamilton's principal function (effectively what we call the action).

Bohm apparently defines the velocity not as an operator on the Hilbert space but as the rate of change $\vec{v}_\psi(t, \vec{x})$ of the so-called configuration $\vec{Q}(t, \vec{x})$ of the system by the relation

$$\vec{v}_\psi = \frac{d\vec{Q}}{dt} = \frac{1}{m}\vec{\nabla}S. \quad (2.57)$$

This rate of change is equal to the classical velocity and not to the quantum velocity and $\vec{Q}$ is the hidden variable we need to adjoin to the wave function $\psi(t, \vec{x})$ in order to obtain a complete description of the system. This definition is also motivated by the definition of the probability current density (see below). In terms of the wave function, Bohm's velocity can then be rewritten as

$$\vec{v}_\psi = \frac{d\vec{Q}}{dt} = \frac{\hbar}{m}\text{Im}\frac{\psi^*\vec{\nabla}\psi}{\psi^*\psi}. \quad (2.58)$$

This form can also be deduced on general grounds by employing symmetry considerations: Galilean invariance (normalization), time reversal (complex conjugation) and rotational invariance (derivation), etc [61].

Thus, the wave function provides the source for Bohm's velocity $\vec{v}_\psi$, which means in particular that Bohm's position of the particle, which is given by $\vec{Q}$, is guided by the wave function and hence the name 'pilot wave' of this interpretation. As said in [61]: 'the wave function governs the evolution of the position of the particle'.

Equations (2.50) and (2.58), where the state of the system is given by the pair $(\psi, \vec{Q})$, define Bohmian (Bohemian) quantum mechanics.

The configuration $\vec{Q}$ lives in a configuration space similar to the configuration space of generalized coordinates found in classical mechanics. Thus $\vec{Q}$, which is a point in a configuration space, is not the same as $\vec{\hat{x}}$, which is an operator on the Hilbert space. However, in Bohemian mechanics what is interpreted as the actual vector position of the quantum particle in ordinary space is in fact $\vec{Q}$ and not the eigenvalue $\vec{x}$ of the vector position operator $\vec{\hat{x}}$. The difference between the two velocities is also exhibited by the fact that since $\vec{Q}$ and $\vec{v}_\psi$ are sourced by the wave functions they must depend on $\vec{x}$ and t. Only in the classical limit do the two, of course, coincide.

We go back now to equation (2.25) and substitute Bohm's velocity. We obtain

$$\frac{\partial S}{\partial t} + \frac{1}{2}m\vec{v}_\psi^2 + V + U = \frac{i\hbar}{2\rho}\left(\frac{\partial \rho}{\partial t} + \vec{v}_\psi \vec{\nabla}\rho + \rho\vec{\nabla}\vec{v}_\psi\right), \quad (2.59)$$

where we have introduced the probability density in the usual way

$$\rho = R^2 = \psi^*\psi, \quad (2.60)$$

and U is the so-called quantum potential defined by

$$U = -\frac{\hbar^2}{2m}\frac{\vec{\nabla}^2 R}{R}. \quad (2.61)$$

Let us also recall the continuity equation (by using the Schrödinger equation)

$$\frac{\partial}{\partial t}(\psi^*\psi) = -\frac{\hbar}{2im}\vec{\nabla}(\psi^*\vec{\nabla}\psi - \psi\vec{\nabla}\psi^*). \quad (2.62)$$

The current is then defined by

$$\vec{J} = \frac{\hbar}{2im}(\psi^*\vec{\nabla}\psi - \psi\vec{\nabla}\psi^*) = \rho\vec{v}_\psi. \quad (2.63)$$

The continuity equation becomes

$$\frac{\partial}{\partial t}\rho = -\vec{\nabla}(\rho\vec{v}_\psi). \quad (2.64)$$

The right-hand side of equation (2.59) is thus equal to the continuity equation and by substitution we also obtain the modified (by the quantum potential) Hamilton–Jacobi equation

$$\frac{\partial S}{\partial t} + \frac{1}{2}m\vec{v}_\psi^2 + V + U = 0 \Rightarrow -\frac{\partial S}{\partial t} = \frac{1}{2m}(\vec{\nabla}S)^2 + V + U. \quad (2.65)$$

This is truly a deterministic formalism since the configuration (the quantum vector position) $\vec{Q} = \int dt \vec{\nabla}S/m$ obeys classical dynamics with an extra piece (the quantum potential U) added to the potential. But it is a non-local formulation since the evolution in time of $\vec{Q}$ is sourced by the wave function ψ which can exist everywhere. The Born rule is imposed here as an initial condition on the wave function.

2.7.2 Beables

Bell is without doubt the most profound thinker of all time on the topic of quantum mechanics. He is the originator of Bell's theorem which remains one of the most fundamental concrete results in the foundation of quantum mechanics, which has also been confirmed experimentally. In this section we will discuss one of his ingenious interpretations of quantum mechanics [62, 63], which is based on Bohm's interpretation of non-relativistic many-particle quantum mechanics [28]. See also [64, 65].

In Bohm's deterministic theory, particles have always definite positions and their motion is fully deterministic guided by the wave function (a pilot wave as Bell called it), which acts as a quantum force rather than as a description of the state of the system.

By analogy, in Bell's indeterministic theory we consider a set of commuting observables (operators or variables) called the beables, which then can be diagonalized simultaneously with simultaneous eigenspaces denoted by S_i called the viable subspaces. In other words, the commuting observables have definite eigenvalues on these eigenspaces so that the actual state of the system is a state vector in one of the viable subspaces S_i.

The evolution of this state is, however, governed by the pilot wave which is a state vector $|\psi(t)\rangle$ obeying the Schrödinger equation with a Hamiltonian given by the physics of the system. This is in direct analogy with the fact that the position of the particles in Bohm's theory (here played by the eigenvalues of the commuting observables) are guided by the Schrödinger wave function. The real state of the system at any given time t is given by one of the components $|\psi_i(t)\rangle$ of the pilot wave, namely

$$|\psi(t)\rangle = \sum_i |\psi_i(t)\rangle. \tag{2.66}$$

Now the real state changes in time stochastically (this is where the indeterministic component enters the formalism) by making transitions between the viable subspaces with transition probabilities given by Bell's postulate, which I will not state explicitly here.

The end result is that the probability $p_i(t)$ that the real state at any time t is $|\psi_i(t)\rangle \in S_i$, if the probability at the initial time $t = 0$ is given by the Born rule, is also given by the quantum mechanical Born rule

$$p_i(t) = \langle \psi_i(t)|\psi_i(t)\rangle.$$

It can also be shown that the above Bell's indeterministic theory reduces in the continuum limit (to be defined) to Bohm's deterministic theory.

Thus we can have a theory in which any chosen set of commuting observables has a definite value yet the results of measurements are given by the probabilities of quantum mechanics.

2.8 On observer-participancy or consciousness

2.8.1 Wigner's friend experiment

The Wigner's friend experiment is one of the most profound gedanken experiments ever devised. It is an extension of the Schrödinger's cat experiment in which the cat is replaced by Wigner's friend. It shows among other things that the collapse of the wave function is a fundamentally different process than unitarity, and in fact collapse cannot be reduced to unitarity, and furthermore it shows that the conscious mind indeed seems to play a genuine real role in measurement.

In some sense collapse is an entirely different interaction in the Universe, a sort of a fifth force so to speak, which occurs after the interaction between the conscious observer and the physical system during the process of quantum measurement.

Wigner's friend experiment can be described as follows. We consider an experimenter F (Wigner's friend) performing an experiment on a two-state quantum system—perhaps a coin C with orthonormal basis states $|\text{heads}\rangle_C$ and $|\text{tails}\rangle_C$. This coin can be replaced by Schrödinger's cat with orthonormal states $|\text{alive}\rangle_C$ and $|\text{dead}\rangle_C$. In Wigner's original version [52] of this experiment this two-state quantum system is given by an object with the states $|\psi\rangle_1$ if a flash emitted by the object has been seen by the friend F, and $|\psi\rangle_2$ if no flash was seen.

The Wigner's friend experiment contains also Wigner W who performs the measurement on the joint system of friend F plus the two-state system. The initial state of the two-state system is assumed to be a linear combination of $|\psi\rangle_1$ and $|\psi\rangle_2$ given by the state vector (assuming the original language of Wigner)

$$|\psi\rangle = \alpha|\psi\rangle_1 + \beta|\psi\rangle_2. \tag{2.67}$$

The complex numbers α and β are the probability amplitudes corresponding to the pure states $|\psi\rangle_1$ and $|\psi\rangle_2$ and their modulus square $|\alpha|^2$ and $|\beta|^2$ give precisely the probabilities of seeing a flash (alive cat, heads) and not seeing a flash (dead cat, tails).

If the state of the object is $|\psi\rangle_1$ then after the interaction between the object and Wigner's friend, which occur during the measurement performed by the friend on the object, the state of their joint system becomes $|\psi\rangle_1 \otimes |F\rangle_1$, where $|F\rangle_1$ is the state of Wigner's friend in which he responds to the question 'Have you seen a flash (dead cat, heads)?' with the answer 'yes'. Similarly, if the state of the object is $|\psi\rangle_2$ then after the measurement performed by the friend on the object the state of their joint system becomes $|\psi\rangle_2 \otimes |F\rangle_2$, where $|F\rangle_2$ is the state of Wigner's friend in which he responds to the above question 'Have you seen a flash (dead cat, heads)?' with the answer 'no'. By the linearity of the Schrödinger equation the joint system friend + object is described by the state vector

$$|\psi\rangle_{\text{joint}} = \alpha|\psi\rangle_1 \times |F\rangle_1 + \beta|\psi\rangle_2 \times |F\rangle_2. \tag{2.68}$$

This is a maximally entangled Bell state [29].

Now, Wigner will perform his measurement on the joint system friend + object. He will ask his friend whether or not he saw a flash (dead cat, heads) and inspect the object. The probabilities according to the Born rule are as follows:
- There is a probability $|\alpha|^2$ that the friend says 'yes' and the object from then on behaves as if it is in the state $|\psi\rangle_1$ of a flash being emitted (or alive cat or heads for the coin).
- There is a probability $|\beta|^2$ that the friend says 'no' and the object from then on behaves as if it is in the state $|\psi\rangle_2$ of a flash not being emitted (or dead cat or tails for the coin).

- There is a probability zero that the friend says 'yes' but the object from then on behaves as if it is in the state $|\psi\rangle_2$ of a flash not being emitted (or dead cat or tails for the coin).
- There is a probability zero that the friend says 'no' but the object from then on behaves as if it is in the state $|\psi\rangle_1$ of a flash being emitted (or alive cat or heads for the coin).

If the corresponding vector states of Wigner in the cases where there is a non-zero probability are denoted by $|F\rangle_{1,2}$ the total state of the joint system friend + object + Wigner is given by the maximally entangled tripartite Greenberger–Horne–Zeilinger (GHZ) state

$$|\psi\rangle_{\text{total}} = \alpha|\psi\rangle_1 \times |F\rangle_1 \times |W\rangle_1 + \beta|\psi\rangle_2 \times |F\rangle_2 \times |W\rangle_2. \qquad (2.69)$$

Everything seems good, but is it really?

If we substitute for Wigner's friend a device, i.e. a measurement apparatus, taken to be just an atom in Wigner's original description, and then repeat the experiment, everything will go through as described above, and indeed nothing can be discerned to be particularly wrong with the above picture.

However, with Wigner's friend instead of the atom performing the measurement on the object, Wigner can simply and surely ask his friend, after completing the experiment, whether or not he saw a flash before he actually had asked him.

It is for certain that the friend will say that he saw the flash or that he did not see the flash, as the case may be, before Wigner asked him. This means in particular that in the reference frame (so to speak) of Wigner's friend the state vector, even before Wigner's measurement, was already either $|\psi\rangle_1 \times |F\rangle_1$ or $|\psi\rangle_2 \times |F\rangle_2$ and not their linear combination, which is in gross contradiction to the above quantum mechanical rules verified experimentally for the atom to a great accuracy.

This is not to say that the friend's position is less reasonable since quantum mechanics assumes him (in the reference frame of Wigner) to occupy the linear combination $|\psi\rangle_{\text{joint}}$ which implies in a clear sense as Wigner puts it: 'that my friend was in a state of suspended animation before he answered my question' [52].

Wigner concludes this experiment by saying [52]: 'It follows that the being with a consciousness must have a different role in quantum mechanics than the inanimate measuring device: the atom considered above. In particular, the quantum mechanical equations of motion cannot be linear if the preceding argument is accepted.'

But are we really confident that the description of the Wigner's friend experiment given above is correct? Another assumption entertained by Wigner himself is 'to assume that the joint system of friend plus object cannot be described by a wave function after the interaction', and that the correct description is given in terms of a density matrix. In other words, we should describe the system by a mixed state instead of a pure state. This also corresponds to the statement that the equation of motion becomes highly non-linear when a measurement by a conscious being is performed. Since as we have already said that the measurement (if it can be called so) carried out by the atom is certainly described by a vector in the Hilbert space, i.e. a pure state. The density matrix can be given by

$$\begin{pmatrix} |\alpha|^2 & \alpha\beta^* \cos\delta \\ \alpha^*\beta \cos\delta & |\beta|^2 \end{pmatrix}. \tag{2.70}$$

Only the case $\delta = 0$ corresponds to orthodox quantum mechanics, i.e. to a pure state, whereas all other states are statistical mixtures with all the properties required by the theory of measurement. The above density matrix defines a continuous transition from a pure state $|\psi\rangle_{\text{joint}}$ to the mixtures $|\psi\rangle_1|F\rangle_1$ and $|\psi\rangle_2|F\rangle_2$.

In summary, this is an objective-collapse model. In general we have [66]:
  i. No-collapse models such as many-worlds and Bohm.
  ii. The objective-collapse model which is a possible view of Wigner himself. Thus, every measurement will produce a collapse for everybody and hence in this case even for Wigner the joint total system is not described by a wave function.
  iii. Subjective-collapse models in which every observer is assigned a collapse in her own measurement only. This is the standard view of Wigner and most of the Copenhagen school. Thus, every measurement will produce a collapse only with respect to the observer performing the measurement.

But does assuming the existence of consciousness and collapse imply any contradictions with physical laws? In other words, do we really have 'a violation of physical laws where consciousness plays a role' as Wigner himself puts it in his article [52]? We do not think this to be the case although Wigner himself has since wavered from his position (see [67] for a brief review). On the contrary, we think that the objective-collapse model can provide a powerful physical handle on consciousness via the interaction between the Universe and the mind. In other words, the Cartesian mind/body problem is not just another metaphysical theory but it can be turned by means of the collapse into a full physical theory. This picture is further strengthened with the established duality between the Copenhagen and many-worlds interpretations, as we will see.

2.8.2 The von Neumann–Wigner interpretation

The Heisenberg cut is a concept introduced by von Neumann to delineate the boundary between the observer and the observed. In classical mechanics the Heisenberg cut is at ∞, thus placing the observed effectively outside the influence of the observer. However, in quantum mechanics its placement is arbitrary.

The von Neumann–Wigner interpretation is an extreme limit of the Copenhagen interpretation in which the Heisenberg cut is placed between the physical brain and the non-physical conscious mind. Hence, the mind is a fundamental entity not reducible to matter, i.e. it is an independent substance, and consciousness is thus a fundamental aspect of nature on equal footing with atoms and elementary particles (as advocated by Chalmers on purely philosophical grounds [68]).

The physical universe is the only true quantum system and the non-physical mind is the only true measuring device. It seems then that the physical world is quantum and the non-physical mind is classical (similar to the many-minds interpretation where the minds are classical as we will discuss shortly) which creates the dichotomy.

The property of unitarity holds true until information enters the conscious mind, i.e. the collapse is caused by the mind performing measurement on brain.

In some sense the physical world does not exist without observing it and the collapse is a real causal interaction between the physical world and the mind. It is in fact a fifth force. The collapse in this interpretation is objective but it is not Penrose's orchestrated objective reduction Orch OR. However, it is the mind that causes the collapse of the wave function (and not the other way around as in Orch OR). Thus, the collapse is the causal link between matter and mind.

Also Cartesian dualism is seen as an intrinsic property of the von Neumann–Wigner interpretation and viewed as a shortcoming. However, quantum dualism is fundamentally different from Cartesian dualism as we will try to argue (see also Stapp, Albert and Lockwood).

Another objection against the von Neumann–Wigner interpretation is due to Bohm and Hiley who say that 'it is difficult to believe that the evolution of the Universe before the appearance of human beings depended fundamentally on the human mind' [69]. Bohm and Hiley themselves provided then, although in a cynical tone, an answer to this same objection by positing a universal mind.

Yet, another answer, is to assume that the mind is formed exactly of the kind of dark matter and/or dark energy which permeates the Universe with real measurable effects without being directly or easily observable. In this way, the mind formed out of this dark stuff is interacting with the brain formed from the luminous stuff, in the same way that dark energy and dark matter found in the Universe interact with ordinary matter. In other words, this dark energy/dark matter is in some sense the universal mind that Bohm and Hiley are positing, and hence the dark stuff in the Universe acts as a universal mind which causes the collapse, upon measurement, to a particular history which we find ourselves observing from the inside.

Another speculative answer to the above objection is to simply deny the existence of the Big Bang (and the corresponding evolution of the Universe) and to say that all history is actually fake, which can be shown by employing the argument of Price against Boltzmann's view concerning the emergence of time from the second law of thermodynamics [31].

It was noted by [70] that the von Neumann–Wigner interpretation can be tested experimentally since if the mind performs a measurement on the brain at time t for the position of a particle, then we can observe the effect of the collapse by measuring the momentum of the same particle at times $t - a$ and $t + a$, i.e. we can measure that the particle does not obey the Schrödinger equation!

2.8.3 The extended Wigner's friend experiment

We consider now the extension of the Wigner's friend experiment outlined recently by Frauchiger and Renner [71]. See also Baumann, Hansen and Wolf [66] and Sudbery [72].

In this experiment we consider Wigner W and his assistant A who perform measurements on Wigner's two friends F_2 and F_1, respectively, who in turn performs their measurements on two two-state quantum systems: an electron spin S in the states $|+\rangle$, $|-\rangle$ and a quantum coin C in the states $|\text{heads}\rangle$, $|\text{tails}\rangle$. The experiment consists in the following:

- We prepare the quantum coin in the state

$$|\psi_0\rangle_C = \frac{1}{\sqrt{3}}|\text{heads}\rangle + \sqrt{\frac{2}{3}}|\text{tails}\rangle. \tag{2.71}$$

- At time t_1 the friend F_1 measures the face of the coin with memory states denoted by $|H\rangle_1$ if she finds heads and $|T\rangle_1$ if she finds tails. The state of the joint system becomes

$$|r\rangle = \frac{1}{\sqrt{3}}|\text{heads}\rangle|H\rangle_1 + \sqrt{\frac{2}{3}}|\text{tails}\rangle|T\rangle_1. \tag{2.72}$$

The friend F_1 sets then the spin of the electron to $|-\rangle$ if she gets heads and to $|+\rangle_x = (|+\rangle+|-\rangle)/\sqrt{2}$ if she gets tails. The joint state of $F_1 + S + C$ is

$$|F_1SC\rangle = \frac{1}{\sqrt{3}}|\text{heads}\rangle|H\rangle_1|-\rangle + \sqrt{\frac{2}{3}}|\text{tails}\rangle|T\rangle_1|+\rangle_x. \tag{2.73}$$

This can be put into the form

$$|F_1SC\rangle = \frac{1}{\sqrt{3}}|\text{tails}\rangle|T\rangle_1|+\rangle + \sqrt{\frac{2}{3}}|\text{fail}\rangle_{1C}|-\rangle, \tag{2.74}$$

where the new state is defined by

$$|\text{fail}\rangle_{F_1C} = \frac{1}{\sqrt{2}}(|\text{heads}\rangle|H\rangle_1 + |\text{tails}\rangle|T\rangle_1). \tag{2.75}$$

The orthogonal state is naturally defined by

$$|\text{ok}\rangle_{F_1C} = \frac{1}{\sqrt{2}}(|\text{heads}\rangle|H\rangle_1 - |\text{tails}\rangle|T\rangle_1). \tag{2.76}$$

- At time t_2 the friend F_2 measures the spin of the electron in the basis $\{|+\rangle, |-\rangle\}$ with corresponding memory states $|U\rangle$ and $|D\rangle$. The system of the joint system $F_2 + F_1 + S + C$ is immediately given by

$$|F_2F_1SC\rangle = \frac{1}{\sqrt{3}}|\text{tails}\rangle|T\rangle_1|+\rangle|U\rangle_2 + \sqrt{\frac{2}{3}}|\text{fail}\rangle_{1C}|-\rangle|D\rangle_2. \tag{2.77}$$

Similarly, we introduce the two states

$$|\text{fail}\rangle_{F_2S} = \frac{1}{\sqrt{2}}(|-\rangle|D\rangle_2 + |+\rangle|U\rangle_2). \tag{2.78}$$

$$|\text{ok}\rangle_{F_2S} = \frac{1}{\sqrt{2}}(|-\rangle|D\rangle_2 - |+\rangle|U\rangle_2). \tag{2.79}$$

Then the state of the joint system $F_2 + F_1 + S + C$ can be put into the form

$$|F_2F_1SC\rangle = \frac{1}{2\sqrt{3}}(|\text{ok}\rangle_1|\text{ok}\rangle_2 - |\text{ok}\rangle_1|\text{fail}\rangle_2 + |\text{fail}\rangle_1|\text{ok}\rangle_2 + 3|\text{fail}\rangle_1|\text{fail}\rangle_2). \quad (2.80)$$

- At time t_3 Wigner's assistant A measures F_1 with the coin in the basis $\{|\text{fail}\rangle_{F_1C}, |\text{ok}\rangle_{F_1C}\}$. The relevant state of the joint system is

$$|a\rangle = \frac{1}{2\sqrt{3}}|\text{ok}\rangle_1(|\text{ok}\rangle_2 - |\text{fail}\rangle_2)|\text{ok}\rangle_a + \frac{1}{2\sqrt{3}}|\text{fail}\rangle_1(|\text{ok}\rangle_2 + 3|\text{fail}\rangle_2)|\text{fail}\rangle_a, \quad (2.81)$$

where $|\text{ok}\rangle_a$ and $|\text{fail}\rangle_a$ are the corresponding memory states. Note that this measurement does not affect the measurement of F_2 in the previous moment since the entangled state of F_2 and S is unchanged.

- At time t_4 Wigner W himself measures F_2 with the spin in the basis $\{|\text{fail}\rangle_{F_2S}, |\text{ok}\rangle_{F_2S}\}$. We obtain the state

$$|w\rangle = \frac{1}{2\sqrt{3}}(|\text{ok}\rangle_1|\text{ok}\rangle_a + |\text{fail}\rangle_1|\text{fail}\rangle_x)|\text{ok}\rangle_2|\text{ok}\rangle_w$$
$$+ \frac{1}{2\sqrt{3}}(-|\text{ok}\rangle_1|\text{ok}\rangle_a + 3|\text{fail}\rangle_1|\text{fail}\rangle_a)|\text{fail}\rangle_2|\text{fail}\rangle_w, \quad (2.82)$$

where $|\text{ok}\rangle_w$ and $|\text{fail}\rangle_w$ are the corresponding memory states. Again we notice that the entangled state of F_1 and C is unchanged and thus the previous measurement of F_1 is not affected. In some limited sense, the measurements of Wigner and his assistant are independent of each other.

- Wigner and his assistant compare their results. If $a = w = \text{ok}$ the experiment stops, otherwise the experiment repeats.

In the above equations we have implicitly assumed standard quantum mechanics, i.e. Copenhagen rules including the projection postulate, which the authors Frauchiger and Renner demanded of any quantum theory to satisfy as a fundamental postulate which they called 'compliance with the quantum theory' (QT). This is the most important among their three postulates. The second postulate which they called 'single-world' (SW) is also very natural even in many-worlds. The third postulate they called 'self-consistency' (SC) in which they required logical consistency between the theory's statements about measurement. We have then
1. QT: compliance with the quantum theory.
2. SW: single-world.
3. SC: self-consistency.

Their main result is the following theorem: 'No physical theory T can satisfy (QT), (SW), and (SC)'.

It is almost obvious that it is the third postulate that can break down and the physics of the relativity of simultaneity provides a very good example. We have no (experimental) doubt that standard quantum theory applies through and through even if the collapse is not a fundamental law but an approximation.

We have now the following results (following the presentation of [72]):
- If the measurement of F_1 at time t_1 gives $r_1=$ 'tails', she will prepare the spin in the state $|+\rangle_x$. The measurement of F_2 at time t_2 puts the system $F_2 + F_1 + S + C$ in the state

$$|F_2F_1SC\rangle = \frac{1}{\sqrt{2}}(|\text{fail}\rangle_1 - |\text{ok}\rangle_1)|\text{fail}\rangle_2. \qquad (2.83)$$

The state of F_2 and S is not affected with the measurement of A at time t_3 of $F_1 + C$. The measurement of W at time t_4 is then given immediately by the result $w_4=$ 'fail'. We have then

$$r_1 = \text{'tails'} \Rightarrow w_4 = \text{'fail'} \leftrightarrow w_4 = \text{'ok'} \Rightarrow r_1 = \text{'heads'}. \qquad (2.84)$$

The last statement is a consequence of the SW postulate.
- If the measurement of F_1 at time t_1 gives $r_1 =$ 'heads', she will prepare the spin in the state $|-\rangle$. The measurement of F_2 at time t_2 gives then the spin $s_2 = -1/2$ with certainty (in the reference frame of F_1). We have then

$$r_1 = \text{'head'} \Rightarrow s_2 = -1/2. \qquad (2.85)$$

- However, in the reference frame of F_2 the full state of the system $F_1 + S + C$ after the F_1 measurement is actually

$$|F_1SC\rangle = \frac{1}{\sqrt{3}}|\text{tails}\rangle|T\rangle_1|+\rangle + \sqrt{\frac{2}{3}}|\text{fail}\rangle_{1C}|-\rangle. \qquad (2.86)$$

Thus, the spin is not $s_2 = -1/2$ with certainty. In the case the measurement of F_2 at time t_2 gives the spin $s_2 = -1/2$, the measurement of A at time t_3 of the joint system $F_1 + C$ will return $a_3 =$ 'fail'. We have then

$$s_2 = -1/2 \Rightarrow a_3 = \text{'fail'}. \qquad (2.87)$$

- From the above three results we have immediately

$$w_4 = \text{'ok'} \Rightarrow r_1 = \text{'head'} \Rightarrow s_2 = -1/2 \Rightarrow a_3 = \text{'fail'}. \qquad (2.88)$$

In other words, the two super-observers Wigner and his assistant can never agree and the procedure will never halt.
- On the other hand, from the above state $|F_2F_1SC\rangle$ it is obvious that the coefficient of $|\text{ok}\rangle_1|\text{ok}\rangle_2$ is non-zero. Hence

$$a_4 = w_4 = \text{ok}, \quad \text{probability} = \frac{1}{12}. \qquad (2.89)$$

The notation a_4 means that if the assistant A performs his experiment at t_4 he will find $a_4 =$ 'ok'. But A performs his measurement at t_3. Fortunately, the measurement of Wigner W at t_4 of $F_2 + S$ does not affect the entangled state of F_1 and C and also it does not affect the entangled state of A and $F_1 + C$. Hence $a_3 = a_4$, i.e. the measurement of A is the same whether he performs it at t_3 or t_4. We have then

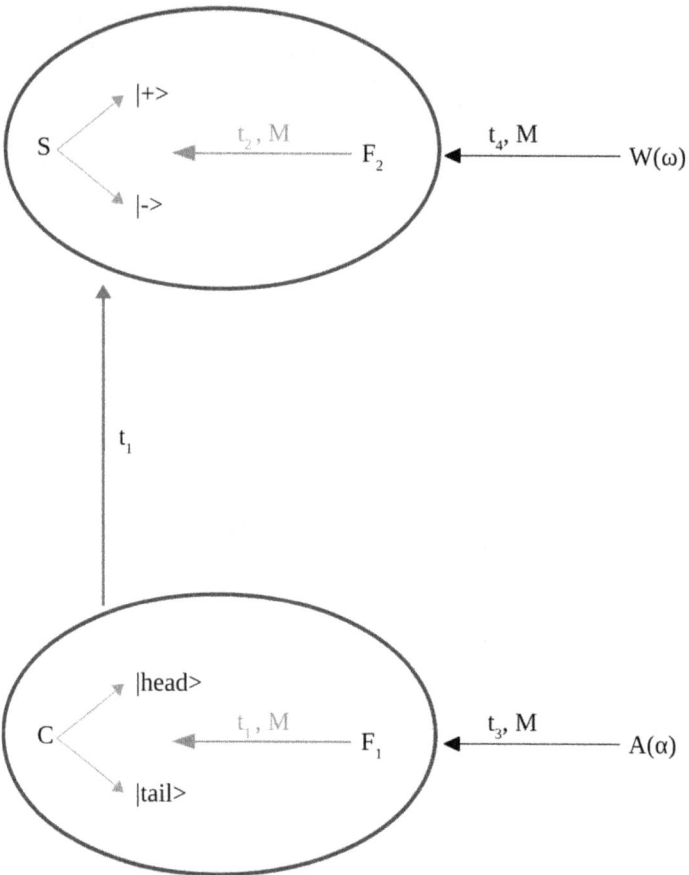

Figure 2.6. The extended Wigner's friend experiment.

$$a_3 = w_4 = \text{ok}, \quad \text{probability} = \frac{1}{12}. \tag{2.90}$$

This is simply in gross contradiction to the previous result.

As pointed out in [66] to make sense of the above contradiction we should differentiate between three models:
  i. No-collapse models.
  ii. Objective-collapse models.
  iii. Subjective-collapse models.

The Wigner–Frauchiger–Renner experiment (figure 2.6) shows in a drastic way that subjective-collapse models are highly untenable in the case of encapsulated observers. The problem lies, as noted by [66], in the questionable possibility of communication between Wigner W and his assistant A. In fact Wigner W and his assistant A play similar roles to the Schwarzschild (external) and infalling (internal) observers in black holes dynamics. Further discussion of this issue, with connection with the ER=EPR conjecture, is delegated until we discuss the ER=EPR conjecture.

2.8.4 The mind/body problem and quantum Zeno effect: Stapp's theory

The quantum Zeno effect plays an essential role in the mind/body theory of Stapp [56] (see also Eccles [73]) which is claimed to avoid the usual problems of classical physicalism and the philosophy of mind given in general by the two problems [74]:

- *The problem of mental causation*: how could the mental cause anything material, particularly in a world which is conceived in a fundamental way as physical substance.
- *The hard problem of consciousness*: how could subjective conscious experience even exist in a physical world dominated by matter and governed by physical laws.

These two problems are solved at once according to Stapp by orthodox quantum mechanics in which (i) the world is not purely physical and perhaps it does not exist independently of conscious observation, (ii) conscious experience is an objective feature of reality and (iii) consciousness is causally efficacious via the collapse postulate.

Stapp asserts rightly that orthodox quantum mechanics is not Cartesian dualism but its intrinsic dualism is physically motivated since the conscious experiences dealt with in quantum mechanics are the core realities of science which consist of 'what we have done and what we have learned' [7]. However, in quantum mechanics and as opposed to physicalism, conscious experience is much more than a mere physical activity. Indeed, events in quantum mechanics are psychophysical events where the state vector describes only their potentialities. This psychophysical event has two aspects: the psychological conscious experience and the physical aspect given by the reduction of the state vector under observation. Thus, consciousness according to quantum mechanics can certainly exist in a physical world which is dynamically and logically complete and also causally closed.

The mind/body problem is then resolved by means of the quantum Zeno effect as follows. Each event, as we have said, is a psychophysical event whose physical aspect is the von Neumann process I, while the associated psychophysical aspect is the conscious experience of intending or choosing to do some physical or mental action. The so-called 'template of action' for some action X is an actualization, via the quantum collapse, of a particular brain pattern and which if held in place for a sufficiently long time will cause the physical action X to occur. Thus, if a sequence of similar process I actions, corresponding to some measurement actions, is repeated consciously in a sufficiently rapid succession, the physical brain state corresponding to that particular template of action will be forced with high probability, and the corresponding physical action will follow. According to Stapp the process I probing actions are freely chosen since they are not determined by the state of the Universe. In summary, mental effort and conscious choice are causally efficacious in the physical world since they can influence and affect the person's physical brain processes in the required way via the quantum Zeno effect.

A specific theory of the brain/mind connection which relies heavily on the quantum Zeno effect in the brain is also proposed by Stapp. In this theory two coupled harmonic oscillators (HOs) are described semi-classically via coherent states. These HOs are modeling the 40 Hz synchronous oscillations of the electromagnetic field at various brain sites which are thought to accompany conscious experience. The first HO describes the brain state of the consciousness of the observer whereas the second HO describes the environment.

If no process I action is initiated then energy is conserved and oscillates back and forth between these two HOs with a period inversely proportional to their coupling. However, if a sequence of process I probing actions is performed in rapid succession faster than the period, then the state of the first HO describing brain correlates of the consciousness of the observer will tend to the uncoupled solution. This is the quantum Zeno effect in this setting. Hence, the brain/mind connection in this theory relies on process I dynamics where the quantum Zeno effect causes the observer's brain to behave in the prescribed way that causes the body to act in accordance with the observer's conscious intent.

For other approaches to the theory of observation in quantum mechanics and the role of the observer see for example [75].

2.9 On observer-determinacy or Orch OR

2.9.1 Orchestrated objective reduction

Schrödinger's cat can be thought of as the most fundamental encapsulation of the so-called measurement problem which is arguably the most central paradox in quantum mechanics. Stated differently, this paradox lies in the profound conflict between von Neumann's processes I and II.

Thus, the state of the cat at time t_0, where t_0 is the time required by an atom before it can decay with a probability amplitude a or not decay with a probability amplitude b, must be given by

$$|\psi(t_0)\rangle = a|\text{decayed}\rangle|\text{dead}\rangle + b|\text{undecayed}\rangle|\text{alive}\rangle. \qquad (2.91)$$

In our case $t_0 = 1$ hour, $a = b = 1/\sqrt{2}$. The state of the cat for later times $t > t_0$ is then given by means of process II as follows:

$$|\psi(t)\rangle = U(t, t_0)|\psi(t_0)\rangle = a(t)|\text{decayed}\rangle|\text{dead}\rangle + b(t)|\text{undecayed}\rangle|\text{alive}\rangle. \qquad (2.92)$$

In other words, the unitary evolution of the quantum state vector $|\psi(t)\rangle$ given by the Schrödinger equation preservers the coherent superposition of the live and dead cats, and in fact it gives the live and dead cats with the same probability since $|a(t)|^2 = |a|^2$ and $|b(t)|^2 = |b|^2$.

But was the cat live or dead in reality before measurement?

According to the Diosi–Penrose–Hameroff interpretation, also known as Orch OR (see [76] and references therein), the resulting density matrix after some time τ is given by a reduced density matrix ρ_r, i.e. if there are N worlds, for example, then in

$N|a|^2$ of them the cat will be indeed alive and in $N|b|^2$ the cat will be dead after τ (a and b do not decrease from their initial values but they simply oscillate in time).

In contrast, in the Copenhagen and the von Neumann–Wigner interpretations the unitary evolution given by the Schrödinger equation will continue indefinitely until and unless an observation is made on the system. Putting it differently, the Diosi–Penrose–Hameroff interpretation assumes that in the unknown theory of quantum gravity (in which spacetime, gravity and matter are unified) a fine-grained unitary evolution must replace the gross-grained Schrödinger evolution which would terminate dynamically.

Also the envisaged quantum structure of spacetime seems to act as a hidden variable in the sense that if the laws of quantum gravity were known to us we would have predicted when the ρ_c (the pre-measurement entangled state) collapses to ρ_r at least on average.

Thus, in Orch OR the system is collapsed long before any measurement has been performed. For Schrödinger's cat the threshold for objective collapse is reached at a time of the order of $\tau = 10^{-43}$ s. This seems to contradict Bell's theorem which shows (or seems to show) that the quantum state does not exist before measurement. Here clearly the state of the cat exists at time $\tau = 10^{-43}$ s as either being dead or being alive long before any measurement has taken place.

In some sense then the envisaged new physics underlying Orch OR acts like hidden variables, which converts the quantum mechanical predictions into a classical probability theory. Thus, the photons emerging from the system (Schrödinger's cat) toward the retina of the observer each carry a classical bit (the cat is either dead or alive in the Orch OR scheme since the superposition has already collapsed).

The brain in Orch OR is assumed to be a quantum system where the basic units of information or qubits are the tubulins with their two conformational states (open and close conformations). The photons arriving at the retina cause a neuronal activity in the brain which constitutes the initial state of this quantum system. The classical computations performed at the level of the neuronal network in the brain and their underlying quantum computations performed by the vast number of tubulins in the vast number of microtubules constituting the cytoskeletons of the vast numbers of neurons provide most of the unconsciousness, pre-consciousness and functional consciousness exhibited by the brain.

The quantum computation proceeds from the initial state via the unitary evolution of the Schrödinger equation and it involves a highly complex superposition of tubulin quantum states which will be resolved by OR. However, the quantum computations at the level of the microtubules performed by the vast number of tubulins, which are isolated against environmental decoherence in the warm, wet and noisy brain (Tegmark), are orchestrated by the entanglement between the tubulins.

These quantum computations must terminate objectively at some time τ by an orchestrated objective reduction (an Orch OR) event which marks one moment of qualitative consciousness (we are seeing, i.e. we are understanding that we are looking at a dead/alive cat).

The neural correlate of the conscious perception emerges then at a time scale of the same order of the lifetime of the coherent superposition before the onset of objective collapse. The continuous flow or stream of consciousness emerges from repeated measurement and OR which hold the conscious perception in place according to the quantum Zeno effect (Stapp).

2.9.2 Time and free will

The proposal that quantum effects do actually occur in microtubules in the brain seems to have been confirmed by the experiments of Bandyopadhyay *et al* [77]. Equally important is the claim that Orch OR solves many outstanding problems in physics and neuroscience. As a very important example we consider briefly the problem of temporal non-locality and free will discovered by Libet [78].

In 1965 Kornhuber (a physician) and Deecke (his student) discovered the famous so-called readiness potential (RP), which is a large and slow potential in the brain (motor cortex) which precedes voluntary movement. Thus it is an event-related potential in the brain. Then in 1983 Libet (a neuroscientist) discovered that the RP begins in the brain almost 350 μs before the conscious decision to move.

Libet asked his experimental subjects to report exactly the time W when they decided to initiate a particular voluntary movement. Then he compared this time W with the time of the onset of the RP and he found that the RP precedes the conscious decision to move by at least 350 μs.

What does this mean?

On the face of it, this simply means that Libet's subjects have consciously and freely decided to initiate a movement only a long time after the preparatory action in the brain has already started. Simply put: the decision to move is not the cause of the motion!

As a consequence, if conscious decisions are not the cause of actions then our conscious free will is not free at all, and it might even be an illusion. This counter-intuitive conclusion led Libet himself (who rejected the obvious and possibly the inevitable consequences for free will) to propose the so-called veto response hypothesis, which states that although our conscious decision is not the cause of the action it can still veto the action before its occurrence.

However, this is not satisfactory at all, and a theory more consistent with this experimental finding is the hypothesis of compatibilism (as opposed to libertarian free will and determinism). This is a very respected and powerful position in the metaphysics of free will. It states that free will and determinism are compatible which is a position adopted by Hume and Leibniz and many other great philosophers.

The veto response hypothesis is effectively a dualistic theory of a libertarian free will executed by a non-deterministic agent. In contrast, in the compatibilist view free will is not absolute freedom but the unfettered ability to act when constraints are absent. In other words, freedom actually relies on the absence of external constraints rather than requiring being outside the causal chain.

The experiments of Libet were repeated by many other groups with the same results and several systematic critiques of such experiments were also put forward. In the end the results stood up to scrutiny.

According to Penrose and Hameroff this effect can be explained naturally by means of the Orch OR. Indeed, since consciousness is a quantum effect the correlation between free will and physical action is not clear-cut but it is in principle fuzzy. Furthermore, quantum entanglement in tubulins by its nature is not expected to be causal in a normal way. Indeed, Orch OR is expected to allow for temporal non-locality and anomalies in the time ordering of consciousness.

References

[1] Feynman R, Leighton B and Sands M 1965 *The Feynman Lectures on Physics* vol 3 (Reading, MA: Addison-Wesley) pp 1.1–8
[2] Susskind L 2016 Copenhagen vs Everett, teleportation, and ER=EPR *Fortsch. Phys.* **64** 551
[3] Susskind L 2016 ER=EPR, GHZ, and the consistency of quantum measurements *Fortsch. Phys.* **64** 72
[4] Wheeler J A 1978 The 'past' and the 'delayed-choice' double-slit experiment *Mathematical Foundations of Quantum Theory* ed A R Marlow (New York: Academic) pp 9–48
[5] Vincent J, Wu E, Grosshans F, Treussart F, Grangier P, Aspect A and Roch J F 2007 Experimental realization of Wheeler's delayed-choice gedanken experiment *Science* **315** 966
[6] Manning A G, Khakimov R I, Dall R G and Truscott A G 2015 Wheeler's delayed-choice gedanken experiment with a single atom *Nat. Phys.* **11** 539–42
[7] Bohr N 1963 *Essays 1958–62 on Atomic Physics and Human Knowledge* (Cambridge: Cambridge University Press)
Bohr N 1949 Discussions with Einstein on epistemological problems in atomic physics *Albert Einstein: Philosopher-Scientist* (Cambridge: Cambridge University Press) see also [11]
[8] von Neumann J 1955 *Mathematical Foundations of Quantum Mechanics* 1st edn (Princeton, NJ: Princeton University Press)
[9] Dirac P A M 1947 *The Principles of Quantum Mechanics* 3rd edn (Oxford: Oxford University Press)
[10] Landau L D and Lifshitz E M 1977 *Quantum Mechanics: Non-relativistic Theory* 3rd edn (Oxford: Pergamon)
[11] Heisenberg W 1958 *Physics and Philosophy* (New York: Harper and Row)
[12] Griffiths R B 2002 *Consistent Quantum Theory* (Cambridge: Cambridge University Press)
[13] Misra B and Sudarshan E C G 1977 The Zeno's paradox in quantum theory *J. Math. Phys.* **18** 756
[14] Russell B 1903 *The Principles of Mathematics* (London: Routledge)
[15] Pascazio S 2013 All you ever wanted to know about the quantum Zeno effect in 70 minutes arXiv: 1311.6645
[16] Patil Y S, Chakram S and Vengalattore M 2015 Measurement induced localization of an ultracold lattice gas *Phys. Rev. Lett.* **115** 140402
[17] Bombelli L, Koul R K, Lee J and Sorkin R D 1986 A quantum source of entropy for black holes *Phys. Rev.* D **34** 3730
[18] Nishioka T, Ryu S and Takayanagi T 2009 Holographic entanglement entropy: an overview *J. Phys.* A **42** 0932
[19] Srednicki M 1993 Entropy and area *Phys. Rev. Lett.* **71** 666

[20] Susskind L and Uglum J 1994 Black hole entropy in canonical quantum gravity and superstring theory *Phys. Rev.* D **50** 2700
[21] Fiola T M, Preskill J, Strominger A and Trivedi S P 1994 Black hole thermodynamics and information loss in two-dimensions *Phys. Rev.* D **50** 3987
[22] Jacobson T 1994 Black hole entropy and induced gravity arXiv: gr-qc/9404039
[23] Solodukhin S N 1995 On 'nongeometric' contribution to the entropy of black hole due to quantum corrections *Phys. Rev.* D **51** 618
[24] Einstein A, Podolsky B and Rosen N 1935 Can quantum mechanical description of physical reality be considered complete? *Phys. Rev.* **47** 777
[25] Bell J S 1966 On the problem of hidden variables in quantum mechanics *Rev. Mod. Phys.* **38** 447, see also [37]
[26] Aspect A, Grangier P and Roger G 1981 Experimental tests of realistic local theories via Bell's theorem *Phys. Rev. Lett.* **47** 460
Aspect A, Grangier P and Roger G 1982 Experimental realization of Einstein–Podolsky–Rosen–Bohm gedanken experiment: a new violation of Bell's inequalities *Phys. Rev. Lett.* **49** 91
Aspect A, Dalibard P and Roger G 1982 Experimental test of Bell's inequalities using time varying analyzers *Phys. Rev. Lett.* **49** 1804
[27] Kochen S and Specker E 1967 The problem of hidden variables in quantum mechanics *J. Math. Mech.* **17** 59
[28] Bohm D 1952 A suggested interpretation of the quantum theory in terms of hidden variables. 1 *Phys. Rev.* **85** 166
Bohm D 1952 A suggested interpretation of the quantum theory in terms of hidden variables. 2 *Phys. Rev.* **85** 180
[29] Bell J S 1964 On the Einstein–Podolsky–Rosen paradox *Physics* **1** 195
[30] Greenberger D M, Horne M A and Zeilinger A 1989 Going beyond Bell's theorem *Bell's Theorem, Quantum Theory, and Conceptions of the Universe* ed M Kafatos (Kluwer: Dordrecht) pp 69–72
[31] Price H 1996 *Time's Arrow and Archimedes Point: New Directions for The Physics of Time* (Oxford: Oxford University Press)
[32] Conway J and Kochen S 2006 The free will theorem *Found. Phys.* **36** 1441
[33] Conway J and Kochen S 2009 The strong free will theorem *Not. AMS* **56** 226–32
[34] Cator E and Landsman K 2014 Constraints on determinism: Bell versus Conway–Kochen *Found. Phys.* **44** 781
[35] Christian W 2010 Can the world be shown to be indeterministic after all? *Probabilities in Physics* ed C Beisbart and S Hartmann (Oxford: Oxford University Press) pp 365–89
[36] Griffiths R B 2011 EPR, Bell, and quantum locality *J. Phys.* **79** 954–65
[37] Bell J S 1987 *Speakable and Unspeakable in Quantum Mechanics. Collected Papers On Quantum Philosophy* (Cambridge: Cambridge University Press) p 212
[38] Stapp H Bell's theorem and world process *Nuovo Cimento* **29B** 270–6
[39] Cramer J 2015 The transactional interpretation of quantum mechanics *Rev. Mod. Phys.* **58** 647–87
[40] Mermin D 1985 Is the Moon there when nobody looks? Reality and the quantum theory *Phys. Today* **38** 38–47
[41] Bohm D 1951 *Quantum Theory* (Englewood Cliffs, NJ: Prentice-Hall)
[42] Griffiths D J 1994 *Introduction to Quantum Mechanics* (Englewood Cliffs, NJ : Prentice-Hall)
[43] Hensen B *et al* 2015 Loophole-free Bell inequality violation using electron spins separated by 1.3 kilometres *Nature* **526** 682

[44] Zeh H D 1970 On the interpretation of measurement in quantum theory *Found. Phys.* **1** 69–76
[45] Zurek W H 1991 Decoherence and the transition from quantum to classical revisited *Phys. Today* **44** 36–44
[46] Tegmark M and Wheeler J A 2001 100 years of the quantum *Sci. Am.* **284** 68
[47] Everett H 1956 Theory of the universal wavefunction *PhD Thesis* Princeton University
[48] Everett H 1957 Relative state formulation of quantum mechanics *Rev. Mod. Phys.* **29** 454–62
[49] DeWitt B S and Graham R N (ed) 1973 *The Many-Worlds Interpretation of Quantum Mechanics* Princeton Series in Physics (Princeton, NJ: Princeton University Press)
[50] Tegmark M 1998 The interpretation of quantum mechanics: many worlds or many words? *Fortsch. Phys.* **46** 855
[51] Wheeler J A 1990 Information, physics, quantum: the search for links *Complexity, Entropy, and the Physics of Information* ed W H Zurek (Redwood City, CA: Addison-Wesley)
[52] Wigner E 2001 Remarks on the mind body question *Philosophical Reflections and Syntheses. The Collected Works of Eugene Paul Wigner. Part B: Historical, Philosophical, and Socio-Political Papers* ed J Mehra vol B/6 (Berlin: Springer)
[53] Wigner E 1963 The problem of measurement *Am. J. Phys.* **31** 6–15
[54] Albert D and Lower B 1988 Interpreting the many worlds interpretation *Synthese* **77** 195–213
[55] Lockwood M 1996 Many minds interpretations of quantum mechanics *Br. J. Phil. Sci.* **47** 159–88
[56] Stapp H P 2009 *Mind, Matter and Quantum Mechanics* (Berlin: Springer)
[57] Schrodinger E 1935 Die gegenwärtige Situation in der Quantenmechanik (The present situation in quantum mechanics) *Naturwissenschaften* **23** 807–12
[58] Trimmer J D 1980 The present situation in quantum mechanics: a translation of Schrödinger's 'cat paradox' paper *Proc. Am. Phil. Soc.* **124** 323–38
[59] Hemmo M and Pitowsky I 2003 Probability and nonlocality in many minds interpretations of quantum mechanics *J. Phil. Sci.* **54** 225–43
[60] De Broglie L 2006 A tentative theory of light quanta *Phil. Mag. Lett.* **86** 411–23
[61] Durr D, Goldstein S and Zanghí N 1992 Quantum equilibrium and the origin of absolute uncertainty *J. Stat. Phys.* **67** 843–907
[62] Bell J S 1982 On the impossible pilot wave *Found. Phys.* **12** 989
[63] Bell J S and Aspect A 2004 Beables for quantum field theory *Speakable and Unspeakable in Quantum Mechanics: Collected Papers on Quantum Philosophy* (Cambridge: Cambridge University Press) pp 173–80
Hiley B and Peat F D 1991 *Quantum Implications Essays in Honour of David Bohm* (London: Routledge)
[64] Sudbery A 1986 *Quantum Mechanics and the Particles of Nature* (Cambridge: Cambridge University Press)
[65] Sudbery A 1987 Objective interpretations of quantum mechanics and the possibility of a deterministic limit *J. Phys. A: Math. Gen.* **20** 1743
[66] Baumann V, Hansen A and Wolf S 2016 The measurement problem is the measurement problem is the measurement problem arXiv: 1611.01111
[67] Esfeld M 1999 Essay review: Wigner's view of physical reality *Stud. Hist. Phil. Mod. Phys.* **30B** 145–54
[68] Chalmers D 1996 *The Conscious Mind: In Search of a Fundamental Theory* (Oxford: Oxford University Press)
[69] Bohm D and Hiley B J 1993 *The Undivided Universe: An Ontological Interpretation of Quantum Theory* (New York: Routledge)

[70] Schreiber Z 1994 The nine lives of Schrödinger's cat arXiv: quant-ph/9501014
[71] Frauchiger D and Renner R 2016 Single-world interpretations of quantum theory cannot be self-consistent arXiv: 1604.07422
[72] Sudbery A 2017 Single-world theory of the extended Wigner's friend experiment *Found. Phys.* **47** 658–69
[73] Eccles J C 1994 *How the Self Controls Its Brain* (Berlin: Springer)
[74] Kim J 2005 *Physicalism or Something Near Enough* (Princeton, NJ: Princeton University Press)
[75] London F and Bauer E 1939 *La Theorie de L'observation en Mecanique Quantique* (Paris: Hermann)
[76] Hameroff S and Penrose R 2014 Consciousness in the Universe: a review of the 'Orch OR' theory *Phys. Life Rev.* **11** 39–78, see also [79]
[77] Sahu S, Ghosh S, Hirata K, Fujita D and Bandyopadhyay A 2013 Multi-level memory-switching properties of a single brain microtubule *Appl. Phys. Lett.* **102** 123701
[78] Libet B 2004 *Mind Time: The Temporal Factor in Consciousness* (Cambridge, MA: Harvard University Press)
[79] Penrose R 1989 *The Emperor's New Mind* (Oxford: Oxford University Press)
Penrose R 2004 *The Road to Reality* (London: Jonathan Cape)

IOP Publishing

Philosophy and the Interpretation of Quantum Physics

Badis Ydri

Chapter 3

The information loss problem in quantum black holes

Toward the goal of drawing systematic parallels between the measurement problem in quantum mechanics and the information loss problem in quantum gravity, we provide in this chapter a comprehensive review of the information loss problem and various theories of quantum gravity. Indeed, in this chapter we provide a description of Hawking radiation and the corresponding information loss problem. Then we review briefly the holographic gauge/gravity duality and many other related ideas relevant to the information loss problem and its unitary resolution, such as the connection between spacetime geometry and quantum entanglement.

3.1 The Schwarzschild black hole

3.1.1 The Schwarzschild black hole and Rindler spacetime

The Minkowski spacetime is arguably the most important metric in physics. It is given in spherical coordinates by

$$ds^2 = -dt^2 + dr^2 + r^2 d\Omega^2. \tag{3.1}$$

The Schwarzschild eternal black hole is the second most important solution of Einstein's equation given by the metric

$$ds^2 = -\left(1 - \frac{2GM}{r}\right)dt^2 + \left(1 - \frac{2GM}{r}\right)^{-1} dr^2 + r^2 d\Omega^2. \tag{3.2}$$

This is the geometry generated by a point mass M placed at the center of spacetime as seen by an observer (Schwarzschild observer) positioned at infinity. It is unique in the sense that any spherical solution of Einstein's equation is necessarily static and thus it must reduce to the Schwarzschild metric (Birkhoff's theorem [1]). The horizon where the coordinate system terminates (the time-like Killing vector vanishes) is located at

$$r_s = 2GM. \tag{3.3}$$

Another important spacetime is Rindler spacetime, which is a uniformly accelerating reference frame with respect to Minkowski spacetime (acceleration a). The near-horizon geometry of a Schwarzschild black hole is in fact a Rindler spacetime with acceleration

$$a = \frac{1}{2r_s}. \tag{3.4}$$

In fact, Rindler spacetime plays, with respect to Minkowski spacetime, the same role that the Schwarzschild observer plays with respect to the so-called Kruskal–Szekeres spacetime, which is the maximally extended Schwarzschild solution.

Hawking temperature from the Unruh effect
The Unruh effect is the statement that a Rindler observer sees the Minkowski vacuum as a thermal canonical ensemble with temperature [2]

$$T = \frac{1}{2\pi}. \tag{3.5}$$

The celebrated Hawking temperature which is seen by the Schwarzschild observer is thus due to gravitational redshift plus the Unruh effect. Indeed, the Rindler time ω is related to the Schwarzschild time t by the relation

$$\omega = \frac{t}{4GM}. \tag{3.6}$$

This leads immediately to the fact that frequency as measured by the Schwarzschild observer ν is redshifted compared to the frequency ν_R measured by the Rindler observer given by

$$\nu_R = 4GM. \, \nu \Rightarrow \nu = \frac{\nu_R}{4GM}. \tag{3.7}$$

Hence the temperature as measured by the Schwarzschild observer is also redshifted as

$$T_R = 4GM \cdot T_H \Rightarrow T_H = \frac{T_R}{4GM} = \frac{1}{8\pi GM}. \tag{3.8}$$

This is precisely the Hawking temperature.

3.1.2 Particle motion in Schwarzschild spacetime

The motion of a scalar particle of energy ν and angular momentum l in the background gravitational field of the Schwarzschild black hole is exactly equivalent to the motion of a quantum particle, i.e. a particle obeying the Schrödinger equation, with energy $E = \nu^2$ in a scattering potential given in the tortoise coordinate r_* with the expression [3]

$$V(r_*) = \frac{r - r_s}{r}\left(\frac{r_s}{r^3} + \frac{l(l+1)}{r^2}\right), \quad r_* = r + r_s \ln\left(\frac{r}{r_s} - 1\right). \tag{3.9}$$

Particles are free both at infinity and at horizon. There is a barrier at $r \sim 3r_s/2$ where the potential reaches its maximum and the height of this barrier is proportional to the square of the angular momentum, namely

$$V_{\max}(r_*) \sim \frac{l^2 + 1}{G^2 M^2} \sim (l^2 + 1)T_H^2, \quad T_H = \frac{1}{8\pi GM}. \tag{3.10}$$

But particles are in thermal equilibrium at the Hawking temperature T_H and thus their energy ν is proportional to T_H. Hence only $l = 0$ particles can escape the potential to infinity (Hawking radiation).

3.1.3 The Kruskal–Szekeres metric and Penrose diagrams

The Kruskal–Szekeres metric, which defines the maximal extension of the Schwarzschild solution, is given by the metric (for T and R given in terms of t and r by some coordinate transformations)

$$ds^2 = \frac{32G^3M^3}{r} \exp\left(-\frac{r}{2GM}\right)(-dT^2 + dR^2) + r^2 d\Omega^2. \tag{3.11}$$

This spacetime consists of four regions separated by future and past horizons.

Region II is the interior of the black hole. Any future directed path in this region will hit the singularity. Region IV is the interior of the so-called white hole. Regions II and IV are called called future and past interiors, respectively. Regions I and III are precisely the exterior regions (asymptotically flat universes) of which we live in one of them.

A white hole is the time reverse of the black hole. This corresponds to a singularity in the past at which the Universe originated. This is a part of spacetime from which observers can escape to reach us while we cannot go there.

We consider now the formation of a black hole from gravitational collapse of a thin spherical shell of massless matter (the red line in figure 3.1). The geometry inside the spherical shell is flat Minkowski spacetime. However, by Birkhoff's theorem the geometry outside the spherical shell is nothing other than Schwarzschild geometry.

The Penrose diagram of Minkowski spacetime is a triangle bounded by five infinities (two light-like infinities, two time-like infinities and one space-like infinity), whereas the Penrose diagram of Schwarzschild geometry consists of the four regions exhibited in the Kruskal–Szekeres metric, which are separated by horizons and end at singularities. Thus the Penrose diagram of the formation of a black hole from gravitational collapse is obtained by gluing the Penrose diagrams of flat Minkowski spacetime (inside the shell) and the Schwarzschild black hole (outside the shell).

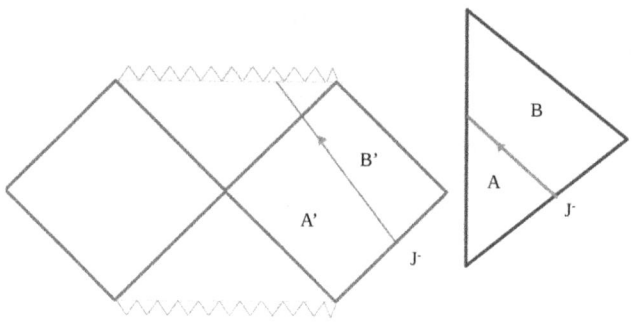

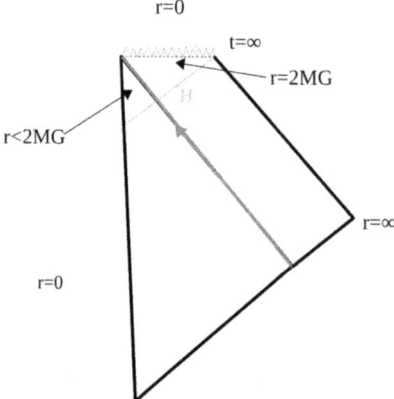

Figure 3.1. The formation of a black hole from gravitational collapse of a thin spherical shell of massless matter (red line).

3.2 Hawking radiation and the information loss problem

3.2.1 Hawking temperature

The Kruskal vacuum state $|0_K\rangle$ plays for a Schwarzschild observer exactly the same role played by the Minkowski state for the Rindler observer, while the Schwarzschild vacuum (also called the Boulware vacuum) plays the same role played by the Rindler vacuum.

The Schwarzschild observer sees (by computing the Bogoliubov coefficients as done, for example, in [4]) the Kruskal vacuum as a heat bath containing n_ω particles given by

$$n_\omega = \langle 0_k | N_\omega | 0_K \rangle = \frac{\delta(0)}{\exp(2\pi\omega/a) - 1}, \quad N_\omega = b_\omega^\dagger b_\omega. \tag{3.12}$$

This a blackbody Planck spectrum with the temperature

$$T_H = \frac{a}{2\pi} = \frac{1}{4\pi r_s} = \frac{1}{8\pi GM}. \tag{3.13}$$

By inserting SI units we obtain

$$T_H = \frac{\hbar c^3}{8\pi GM k_B}. \tag{3.14}$$

The black hole as seen by a distant observer is radiating energy, thus its mass decreases, and as a consequence its temperature increases, i.e. the black hole becomes hotter, which indicates a negative specific heat.

3.2.2 Thermodynamics

In summary, to a distant observer the Schwarzschild black hole appears as a body with energy given by its mass M and a temperature T given by the Hawking temperature

$$T = \frac{1}{8\pi GM}.$$

The thermodynamic entropy S is related to the energy and the temperature by the formula $dU = TdS$. Thus we obtain for the black hole the entropy

$$dS = \frac{dM}{T} = 8\pi GM dM \Rightarrow S = 4\pi GM^2. \tag{3.15}$$

However, the radius of the event horizon of the Schwarzschild black hole is $r_s = 2MG$, and thus the area of the event horizon (which is a sphere) is

$$A = 4\pi(2MG)^2. \tag{3.16}$$

By dividing the above two equations we obtain

$$S = \frac{A}{4G}. \tag{3.17}$$

The entropy of the black hole is proportional to its area. This is the famous Bekenstein–Hawking entropy formula [5, 6].

The last point of primary importance concerns black hole thermodynamics. The thermal entropy is the maximum amount of information contained in the black hole. The entropy is mostly localized near the horizon, but quantum field theory (QFT) gives a divergent value, instead of the above Bekenstein–Hawking value. The number of accessible quantum microscopic states is determined by this entropy via the formula

$$n = \exp(S). \tag{3.18}$$

Since QFT gives a divergent entropy instead of the Bekenstein–Hawking value it must be replaced by quantum gravity (QG) near the horizon and this separation of the QFT and QG degrees of freedom can be implemented by the stretched horizon, which is a time-like membrane, at a distance of one Planck length $l_P = \sqrt{G\hbar}$ from the actual horizon, and where the proper temperature becomes very large and most of the black hole entropy accumulates.

3.2.3 Gravitational collapse

We consider following [7] the more realistic situation in which a Schwarzschild black hole is formed by gravitational collapse of a thin mass shell, as in the Penrose diagram shown in figure 3.1. The vacuum state before collapse is a Minkowski state whereas after collapse it is a Kruskal state.

The initial/incoming state of the black hole is a pure state $|\psi\rangle$ (with incoming negative frequency modes). The final/outgoing state corresponds to a positive frequency mode P, i.e. centered around some positive frequency ω, with support only at large radii r at late times $t \longrightarrow +\infty$.

In order to find the initial state we run P backwards in time toward the black hole. A reflected part R will scatter off the black hole and return to large radii and a transmitted part T with support only immediately outside the event horizon, i.e. $P = R + T$. At the event horizon both positive and negative frequency modes will be seen in T by a freely falling observer. Since the reflected wave packet R only has support in the asymptotic flat region and since $|\psi\rangle$ contains no positive frequency incoming excitations we have

$$a(R)|\psi\rangle = 0. \tag{3.19}$$

Also the black hole state $|\psi\rangle$ does not contain positive high frequency modes throughout (the adiabatic principle). We have then

$$a(T^+)|\psi\rangle = a(\bar{T}^-)|\psi\rangle = 0. \tag{3.20}$$

These equations define the so-called Unruh vacuum $|U\rangle$.

If the state $|\psi\rangle$ were annihilated exactly by T it would have been identical with the Schwarzschild or Boulware state $|B\rangle$. Recall that the Schwarzschild vacuum $|B\rangle$ is annihilated by $a(R)$, $a(T)$ and $a(\tilde{T})$, where $\tilde{T}$ is defined inside the horizon.

The solution of equations (3.20) gives a maximally entangled state describing pairs of particles with zero Killing energy each, namely

$$|U\rangle \sim \sum_n \exp\left(-\frac{n\pi\omega}{a}\right)|n_R\rangle|n_L\rangle. \tag{3.21}$$

The states $|n_R\rangle$ and $|n_L\rangle$ are the level n-excitations of the exterior modes T and the interior modes $\tilde{T}$. Thus one of the pair $|n_R\rangle$ goes outside the horizon (Hawking radiation) whereas the other pair $|n_L\rangle$ falls behind the horizon (information lost).

However, the asymptotic Schwarzschild observer registers a thermal mixed state with Hawking temperature, namely

$$\rho_R = \text{Tr}_L |U\rangle\langle U| \sim \sum_n \exp\left(-\frac{2n\pi\omega}{a}\right)|n_R\rangle\langle n_R|. \tag{3.22}$$

In summary, a correlated entangled pure state near the horizon gives rise to a thermal mixed state outside the horizon. This is the information loss problem.

3.2.4 Information loss problem and the laws of physics

In gravitational collapse an entangled pair of particles is created on the past null infinity $\mathcal{J}^-$ at $r = \infty$ then scatters off the black hole. Then one of the particles escapes to the future null infinity $\mathcal{J}^+$ at $r = \infty$ whereas the other one passes through the horizon and reaches the singularity S at $r = 0$.

From the assumption of locality we have

$$H_{in} = H_-, \quad H_{out} = H_+ \otimes H_S. \tag{3.23}$$

From the assumption of unitarity we have

$$|\psi_{out}\rangle = S|\psi_{in}\rangle. \tag{3.24}$$

However, with respect to the outside observer at the infinity $\mathcal{J}^+$ the final state can only be given by a reduced density matrix since she cannot access the states on the singularity S. We have then

$$|\psi_{in}\rangle\langle\psi_{in}| \longrightarrow \rho_{out} = \text{Tr}_S |\psi_{out}\rangle\langle\psi_{out}|. \tag{3.25}$$

This is the information loss paradox [8, 9].

Thus black hole dynamics contradicts one of the following laws of physics (called laws of nature by Susskind):

- *Information conservation (quantum mechanics)*: The information is defined as the difference between the coarse-grained Boltzmann thermodynamic entropy and the fine-grained von Neumann entanglement entropy. This quantity is conserved due to the unitarity of quantum mechanics.
- *Equivalence principle (general relativity)*: Spacetime is a manifold which is locally Minkowski flat (special relativity) everywhere.
- *Quantum xerox principle (information theory)*: The linearity of quantum mechanics forbids the duplication of quantum information.
- *The Bekenstein–Hawking entropy formula* [5, 6]:

$$S = \frac{A}{4G} = \ln n.$$

3.2.5 Unitarity and the Page curve

The black hole starts in a pure state and after its complete evaporation the Hawking radiation is also in a pure state. This is the assumption of unitarity. The information is given by the difference between the thermal entropy of Boltzmann and the entanglement entropy of von Neumann, whereas the entanglement entropy is given by the von Neumann entropy.

The so-called Page curve [10, 11] is expected to give an entanglement entropy which starts at a zero value, then reaches a maximum value at the so-called Page time and then drops to zero again. The Page time is the time at which the black hole evaporates around one-half of its mass and the information starts to get out with the radiation. Before the Page time only energy gets out with the radiation, while only at the Page time does the information start to get out, and it gets out completely at the moment of evaporation. This is guaranteed to happen because of the second principle of thermodynamics and the assumption of unitarity (figure 3.2).

The computation of the Page curve starting from first principles will provide, in some precise sense, the mathematical solution of the black hole information loss problem (Strominger as recounted by Harlow).

3.3 Black hole complementarity

Two properties are said to be dual or complementary if they cannot be observed simultaneously. Black hole complementarity, due to 't Hooft [12, 13] and Susskind [14], is a prime example. It states simply that no single observer will ever witness a violation of a law of physics.

Thus, the external observer sees the infalling matter heating up the stretched horizon (not the event horizon but a physical membrane above it) which then reradiates back the infalling information in the form of Hawking radiation. The information is uniformly spread out over the stretched horizon. The infalling observer, in contrast, sees (because of the equivalence principle) that there is nothing special happening at the horizon and the infalling information or matter will actually pass through and reach the singularity. The information is seen by this observer as localized on the stretched horizon.

A better statement is to say that the information is both reflected at (with respect to the external observer) and passed through (with respect to the infalling observer) the event horizon. These two stories are in fact complementary not contradictory. The two observers cannot communicate (infinite time dilation at the horizon, no signal can be sent from behind the horizon).

In some sense the complementarity principle resolves the violation of the quantum xerox or no-cloning theorem. Thus, information is either inside the black hole (with respect to an infalling observer) or outside the black hole (with respect to an external observer).

The postulates of black hole complementarity are (external observer):
1. Unitarity (quantum mechanics).
2. Semi-classical field approximation (stretched horizon).
3. The Hawking–Bekenstein law: The number of microstates of a black hole of mass M is equal to the exponential of the entropy $S(M) = A(M)/4G$ where $A(M)$ is the surface area of the black hole.

The conservation of information (no-cloning or xerox theorem) and the equivalence principle are also implicitly assumed.

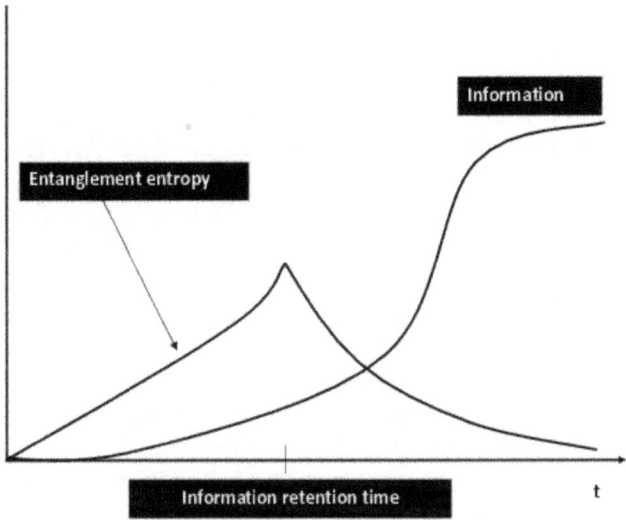

Figure 3.2. Page time, the entanglement entropy and the information.

3.4 The firewall and monogamy

3.4.1 Monogamy

Hawking radiation is formed of pairs of entangled particles created around the horizon with one of each pair passing through the horizon whereas the other one goes to infinity. The final state of the particle outside after complete evaporation of the black hole is mixed (information is lost). The final state must become purified (unitarity).

By using the black hole complementarity we know that the infalling observer sees the bits of information going inside the black hole whereas the external observer sees them in the late Hawking radiation. However, the infalling observer sees the horizon as flat spacetime whereas the external observer sees it as a hot membrane (strong non-locality).

In conclusion, a particle B from the late Hawking radiation must be entangled with the earlier part R_B of Hawking radiation (unitarity). But this particle B must also be entangled with a particle A inside the horizon (locality). As it turns out, this contradicts another cherished principle, which is the principle of monogamy of quantum entanglement: the same system cannot be entangled with two other different systems at the same time.

3.4.2 Firewall

In summary, we have to give either:
1. *Unitarity (purity)*: Black hole information is carried out with the late Hawking radiation. This effect is measured by an external observer.
2. *Effective field theory (EFT)*: Quantum field theory in curved backgrounds and locality must hold. In other words, we suppose that semi-classical gravity

is valid outside the horizon, i.e. small curvature near and outside the horizon. Thus, the external observer will see the horizon as a physical membrane (stretched horizon).
3. *Equivalence principle (no-drama)*: Nothing special happens as we fall across the horizon. This effect is measured by an infalling observer.

The AMPS firewall paradox [15] in our view is the most precise formulation of the information loss paradox. It states that black hole complementarity (first and second postulates) are inconsistent with the equivalence principle (third postulate). It was expected (according to the AMPS) that EFT (locality) breaks down and complementarity is then sufficient. However, the situation seems more complicated.

The firewall is a release of high energy at the horizon arising from breaking the entanglement between the inside and outside modes in accordance with monogamy. This results from the following contradictory statements:
- First, from point 1 above, any particle of the late Hawking radiation (after Page time) is maximally entangled with some subsystem of the early radiation because the early particles have many more states. It is an eigenstate of $bb^\dagger$.
- However, by assumption 2 above, we can propagate (essentially freely) any late Hawking mode back from infinity (where the unitarity assumption should hold) to the horizon (where the no-drama assumption is supposed to hold). The mode becomes highly blue shifted, i.e. it has a very high energy. In other words, we obtain an early mode of the Hawking radiation with very high energy which is encountered by the infalling observer.
- However, from assumption 3 above we conclude that this same Hawking mode is maximally entangled with a mode inside the horizon. The field is in the ground state $a_\omega^\dagger a_\omega = 0$ and the mode is not an eigenstate of $bb^\dagger$.

3.5 ER=EPR and the modification of quantum mechanics
3.5.1 ER=EPR

The ER=EPR proposal [16] (see also [17, 18]) is a much more fundamental conjecture which states that general relativity (Einstein–Rosen or ER wormhole [19]) is equivalent to quantum mechanics (Einstein–Podolsky–Rosen or EPR entanglement [20]). This profound idea is inspired by van Raamsdonk's observation [21] (see also [22]) that a maximally extended AdS black hole is dual via AdS/CFT correspondence to a pair of maximally entangled thermal conformal field theories living at the boundaries [23].

Thus according to the ER=EPR conjecture there is no need for the firewall, which results from the violation of the equivalence principle through the breaking of the quantum entanglement between the interior and the exterior of the black hole, since the spacetime manifold is genuinely smooth across the horizon. Indeed, it is argued that maximal entanglement between different spacetime regions (such as the interior and the exterior of the black hole) acts exactly as a smooth physical bridge between those regions and thus no contradiction results with the monogamy principle. The ER=EPR conjecture entails therefore some substantial modification of quantum mechanics by re-interpreting quantum entanglement as physical gravitational bridges.

The ER bridge or wormhole is a solution of Einstein's equations which connects spacetime points which are very far away. For example in the Schwarzschild spacetime (Kruskal–Szekeres coordinates) the two asymptotically flat universes (regions I and III) are connected by a non-traversable Einstein–Rosen bridge (a wormhole). This ER bridge is, however, non-traversable (locality is maintained) and the two observers can only meet inside the horizon rather than communicate.

The ER bridge generates two maximally entangled regions by Hawking particles. For example the Hartle–Hawking AdS ER bridge yields the maximally entangled state [23, 24]

$$|\psi\rangle \propto \sum_i \exp(-\beta E_i/2)|i^*\rangle_L |i\rangle_R. \tag{3.26}$$

We recall that maximal entanglement is described by Bell states such as EPR pairs and thus locality is maintained since EPR correlations cannot be used to send information faster than the speed of light, corresponding to the fact that the ER bridge is generally a non-traversable Lorentzian wormhole.

Conversely, two distant black holes connected through the interior via an ER bridge can be interpreted as maximally entangled states of two black holes that form an EPR pair (figure 3.3).

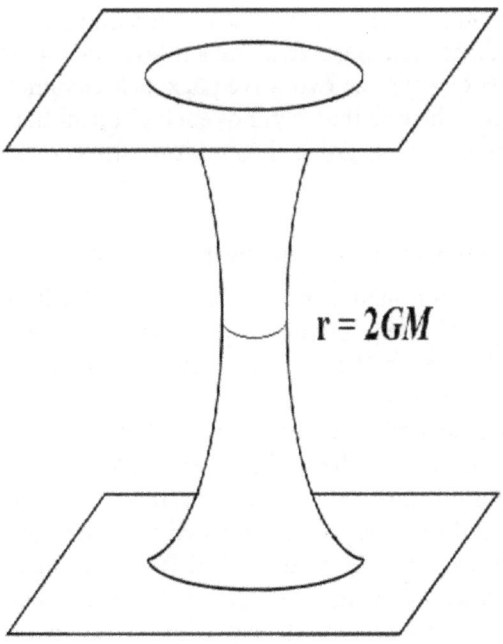

Figure 3.3. Two asymptotically flat universes are connected by a non-traversable ER bridge, i.e. the two observers can only meet inside the horizon.

In summary, every EPR correlated state is connected by an ER bridge (not necessarily the one given by general relativity). The ER=EPR conjecture resolves the information loss problem by stating that the Hawking radiation is actually connected by ER bridges to itself and also to the black hole horizon.

A straightforward generalization of the ER = EPR conjecture, based on the holographic gauge/gravity duality or the AdS/CFT correspondence [25] and the remarkable Ryu–Takayanagi formula [26, 27] which generalizes the Bekenstein–Hawking entropy formula, is the idea that spacetime geometry is equivalent to quantum entanglement [28–31].

3.5.2 Interference as entanglement

As an application of the ER=EPR conjecture we can argue (following Susskind) that entanglement is more fundamental than interference by showing how to reduce the latter to the former.

Every single electron emerges from the two slits in a linear superposition of two non-overlapping wave packets with some relative phase. The wave packets will in time overlap giving rise to the interference pattern. We treat the wave function as a field (second quantization) with the main degrees of freedom of the field found inside the boxes centered around the peaks of the two emerging wave packets. The state of the system is a maximally entangled Bell pair, i.e.

$$|\psi\rangle = \frac{1}{\sqrt{2}}(|10\rangle - |01\rangle). \tag{3.27}$$

The state $|10\rangle$ corresponds to a particle in the left box and no particle in the right box whereas $|01\rangle$ corresponds to no particle in the left box and a particle in the right box.

According to the ER=EPR the two wave packets are connected by an ER bridge or a wormhole which is the one that reminds each electron that the other electron is there. That is how the electron going through one slit knows that the other slit is open or closed

3.5.3 The extended Wigner's friend experiment revisited

The Wigner's friend experiment can also be discussed within the context of the ER=EPR conjecture [17, 18]. Here we propose a resolution of the apparent contradiction found earlier in the context of the extended Wigner's friend experiment by using the ER=EPR conjecture.

The first set of encapsulated observers consists of the assistant A, the friend F_1 and the coin C, which can all be collapsed into entangled black holes (fungibility of entanglement). Communication between various observers is described by entanglement. The friend F_1 and the coin C are described by a Bell state which is equivalent to an EPR bridge or a wormhole by the ER=EPR conjecture.

This bridge is snipped when the assistant makes his measurement and thus no messages can be sent between the friend F_1 and the coin to meet inside the wormhole. However, with respect to a many-worlds description, the assistant A, the friend F_1 and the coin C are described by a tripartite GHZ state which is geometrically equivalent to something called the GHZ brane by the ER=EPR conjecture. By the separability condition satisfied by GHZ states no entanglement exists between any two parties and thus no messages can be sent between any two parties to meet inside the GHZ brane. However, the many-worlds formalism preserves entanglement and thus reversibility. In particular, the union of any two parties in the GHZ state is still entangled with the third party, and hence messages can be sent between any two cooperating parties (A and F_1) and the third one (C) to meet inside the GHZ brane.

The same picture holds for the second set of encapsulated observers consisting of Wigner himself W, the friend F_2 and the spin S, with messages between any two cooperating parties (Wigner and F_2) and the third one (S) can be sent to meet inside their corresponding GHZ brane. There is no direct entanglement between Wigner and his assistant and hence their only route of exchanging messages is through the entanglement between the coin C and the spin S, which in turn entangle the two GHZ branes corresponding to the two sets of encapsulated observers.

With respect to Wigner (or his assistant A), his own measurement has already severed the above structure at the level of the EPR bridge relating the friend F_1 and the coin C (or the friend F_2 and the spin S). Thus the possibility of communication is indeed quite questionable in the case of encapsulated observers.

3.6 The holographic gauge/gravity duality

3.6.1 Holography

The number of degrees of freedom, or equivalently the amount of information, contained in a quantum system is measured, as we know, by thermodynamic entropy. In quantum mechanics the entropy is an extensive quantity and thus the entropy of a d-dimensional spatial region $\mathbf{R}^d$ is proportional to its volume V_d.

However, in quantum gravity the entropy is sub-extensive, i.e. the entropy of a d-dimensional spatial region $\mathbf{R}^d$ is actually proportional to the surface area $S_{d-1} = \partial V_d$ which bounds its volume V_d and is not proportional to the volume V_d itself.

In other words, the entropy of a d-dimensional spatial region, in a gravitational theory, is bounded by the entropy of the black hole which fits inside that spatial region. This is essentially what is called the holographic principle introduced first by 't Hooft [32] and then extended to string theory by Susskind [33] (see also [34]). As we can see this principle is largely inspired by the Bekenstein–Hawking formula which states that the entropy of a black hole S_{BH} is proportional to the surface area A_H of the black hole horizon with the constant of proportionality equal $1/4G_N$ where G_N is Newton's constant, namely

$$S_{\text{BH}} = \frac{A_H}{4G_N}. \tag{3.28}$$

The holographic principle provides therefore a partial answer to the question of how could a higher dimensional gravity theory (AdS_{d+1}) contain the same number of degrees of freedom, the same amount information, and have the same entropy as a lower dimensional quantum field theory (CFT$_d$), i.e. it lies at the heart of the celebrated AdS/CFT correspondence [25].

3.6.2 The black p-branes and supersymmetric Yang–Mills gauge theory

Maldacena in [25] (see also [35, 36]) produced perhaps the greatest feat in theoretical physics since the days of the quantum revolution of Bohr and the relativity revolution of Einstein. Indeed, one can argue that this achievement is even more astounding and even more impressive than the achievement of the standard model of particle physics and the standard model of cosmology.

But what is the gauge/gravity duality?

This can be explained in two axes:

1. First, we consider the maximally supersymmetric $U(N)$ gauge field theory in $p + 1$ dimensions. The action principle of this theory is invariant under the maximum number of supersymmetric transformations in the given dimension (recall that supersymmetry is a generalization of translations and rotations by adding odd Grassmannian coordinates to spacetime). The gauge theory $U(N)$ is a generalization of electromagnetism, i.e. of $U(1)$, to the case of N charges and N anti-charges called colors.

    We consider this theory in the 't Hooft limit given by taking the gauge coupling constant g^2 to zero and N to infinity in such a way that we keep the so-called 't Hooft coupling λ constant, where [37]

    $$\lambda = g^2 \cdot N. \tag{3.29}$$

    It is well known that the above gauge theory in the 't Hooft limit describes N coincident Dirichlet(p)-branes (or Dp-branes for short) which are, in the words of Polchinski, '…extended objects defined by mixed Dirichlet–Neumann boundary conditions in string theory [which] break half of the supersymmetries of the type II superstring and carry a complete set of electric and magnetic Ramond–Ramond charges …' [38].

2. Second, the system of N coincident Dp-branes form something we call the black p-brane [39–41] which is a generalization of black holes. This black p-brane has mass and as such it curves the spacetime around it with a corresponding metric given by a type II supergravity solution. This near extremal p-brane solution is given explicitly by (with $\alpha' = l_s^2 \to 0$, where l_s is the string fundamental length)

$$ds^2 = \alpha' \left[\frac{U^{\frac{7-p}{2}}}{g_{YM}\sqrt{d_p N}} \left(-\left(1 - \frac{U_0^{7-p}}{U^{7-p}}\right) dt^2 + dy_{\parallel}^2 \right) + \frac{g_{YM}\sqrt{d_p N}}{U^{\frac{7-p}{2}}} \frac{dU^2}{\left(1 - \frac{U_0^{7-p}}{U^{7-p}}\right)} \right. \tag{3.30}$$

$$\left. + g_{YM}\sqrt{d_p N}\, U^{\frac{p-3}{2}} d\Omega_{8-p}^2 \right].$$

$(t, y_\parallel)$ are the coordinates along the p-brane world volume, Ω_{8-p} is the transverse solid angle associated with the transverse radius U, whereas U_0 corresponds to the radius of the horizon, and it is given in terms of the energy density of the brane E by

$$U_0^{7-p} = a_p g_{YM}^4 E. \tag{3.31}$$

The string coupling constant g_s in this limit is given by

$$g_s = (2\pi)^{2-p} g_{YM}^2 \left(\frac{d_p g_{YM}^2 N}{U^{7-p}} \right)^{\frac{3-p}{4}}. \tag{3.32}$$

The factors d_p and a_p in the above equations are some numerical constants which depend only on the dimension p. The corresponding Hawking temperature can be determined from the conical singularity of the Euclidean metric to be given by

$$T = \frac{(7-p) U_0^{\frac{5-p}{2}}}{4\pi g_{YM} \sqrt{d_p N}}. \tag{3.33}$$

The Bekenstein–Hawking formula $S = A/4$ where A is the area density of the horizon can also be calculated for this metric from the first law of thermodynamics $dS = dE/T$.

The gauge/gravity duality states simply that the above maximally supersymmetric $U(N)$ gauge theory in $p + 1$ dimensions in the 't Hooft limit is exactly equivalent to type II supergravity in ten dimensions around the above black p-brane. Thus, 'hidden within every non-abelian gauge theory, even within the weak and strong nuclear interactions, is a theory of quantum gravity' [42].

In other words, a higher dimensional curved spacetime manifold emerges in this duality from a lower dimensional gauge theory in a flat spacetime manifold. The emerging extra spatial dimensions are described in the gauge theory by adjoint scalar fields given by $N \times N$ matrices. The extra dimensions emerge in the gauge theory precisely in the limit $N \longrightarrow \infty$, whereas strongly quantum gauge fields give rise to effective classical gravitational fields in the limit $\lambda \longrightarrow \infty$.

This duality provides therefore a very concrete non-perturbative definition of quantum gravity and its quantum geometry in terms of gauge theory. Indeed, since the gauge theory is fully defined quantum mechanically, and in fact it is fully defined non-perturbatively on the lattice a la Wilson [43], the gauge/gravity duality gives then a full non-perturbative definition of quantized gravity which is the holy grail of theoretical physics.

Quantum corrections to the gravity side are given by string theory loop expansion, i.e. in the string coupling constant g_s. This corresponds on the gauge side to $1/N^2$ corrections, whereas stringy corrections to the gravity side are given by the fundamental string length l_s and they correspond on the gauge side to $1/\lambda$ corrections where λ is the above 't Hooft coupling.

The most important examples of gauge/gravity duality are:
1. The celebrated AdS$_5$/CFT$_4$ on AdS$_5 \times S^5$ in $(3 + 1)$ dimension which is relevant to the D3-brane [25].
2. The ABJM theory in $(2 + 1)$ dimension on AdS$_4 \times S^7$ relevant to the M2-brane [44]. The dual gauge theory relevant to the M5-brane (which is the magnetic dual of the M2-brane) on AdS$_7 \times S^4$ is still largely unknown.
3. Matrix string theory in $(1 + 1)$-dimension [45] relevant to the Gregory–Laflamme instability [46] (for example a black string breaking into black holes).
4. The BFSS model or M-(atrix) theory in $(0 + 1)$ dimension [47] which is perhaps the most important case. Indeed, the discrete light cone quantization (DLCQ) of M-theory should be described by the BFSS model. The compactification of the BFSS model on a circle (high temperature limit) gives a $U(N)$ gauge theory in $(0 + 0)$-dimension known as the type IIB matrix model (also known as the IKKT model), which provides a non-perturbative regularization of type IIB superstring theory in the Schild gauge [48].
5. The BMN model in $(0 + 1)$ dimension [49], which is a simple mass deformation of the BFSS model describing the M2-brane and the M5-brane in the Penrose limit of the AdS$_4 \times S^7$ and AdS$_7 \times S^4$.

An overview of these theories from the lattice point of view is given in [50].

In summary, we have:
- The gauge theory in the limit $N \longrightarrow \infty$ (where extra dimensions will emerge) and $\lambda \longrightarrow \infty$ (where strongly quantum gauge fields give rise to effective classical gravitational fields) should be equivalent to classical type II supergravity around the p-brane spacetime.
- The gauge theory with $1/N^2$ corrections should correspond to quantum loop corrections, i.e. corrections in g_s, in the gravity/string side.
- The gauge theory with $1/\lambda$ corrections should correspond to stringy corrections, i.e. corrections in l_s, corresponding to the fact that the degrees of freedom in the gravity/string side are really strings and not point particles.

3.6.3 AdS/CFT correspondence

The near-horizon geometry of the black p-branes involves always the product of spheres and AdS spaces which are maximally symmetric spaces given by the coset spaces $S^{d+1} = SO(d+2)/SO(d+1)$ and AdS$_{d+1} = SO(d,2)/SO(d,1)$ respectively. The dual gauge theory is always a conformal field theory CFT$_d$ living on the boundary of AdS$_{d+1}$.

The Klein–Gordon equation in AdS$_{d+1}$ reads explicitly

$$z^{d+1}\partial_z(z^{1-d}\partial_z\phi) + z^2\partial^2\phi - m^2L^2\phi = 0. \tag{3.34}$$

Let $f_k(z)$ be the Fourier transform in the x-space. Near the conformal boundary $z = 0$ we have the behavior

$$f_k(z) \longrightarrow A(k)z^{d-\Delta} + B(k)z^{\Delta}, \quad z \longrightarrow 0. \tag{3.35}$$

The exponent Δ is the scaling dimension of the field given by

$$\Delta = \frac{d}{2} + \sqrt{\frac{d^2}{4} + m^2 L^2}. \tag{3.36}$$

For $m^2 > -d^2/4L^2$ we have $d - \Delta \leq \Delta$ and hence $z^{d-\Delta}$ is dominant as $z \longrightarrow 0$. The behavior on the boundary is then

$$\phi(z = \epsilon, x) = A(x)\epsilon^{d-\Delta}, \quad \epsilon \longrightarrow 0. \tag{3.37}$$

This is divergent for $m^2 > 0$ and hence the scalar field living on the boundary is identified with $A(x)$, namely

$$\varphi(x) = \lim_{\epsilon \to 0} \epsilon^{\Delta - d} \phi(\epsilon, x). \tag{3.38}$$

This is the scalar field representing (or dual to) the anti-de Sitter scalar field $\phi(z, x)$ at the boundary $z = 0$.

Let $\mathcal{O}(z, x)$ and $\mathcal{O}(x)$ be the dual operators of the scalar fields $\phi(z, x)$ and $\varphi(x)$, respectively, i.e.

$$S_{\text{bound}} = \int d^d x \sqrt{\gamma} \phi(\epsilon, x) \mathcal{O}(\epsilon, x) = L^d \int d^d x \varphi(x) \mathcal{O}(x). \tag{3.39}$$

We obtain

$$\mathcal{O}(\epsilon, x) = \epsilon^{\Delta} \mathcal{O}(x). \tag{3.40}$$

This shows explicitly that Δ is the scaling dimension of the dual operator $\mathcal{O}$.

We are interested in computing the boundary CFT correlation functions

$$\langle \mathcal{O}(x_1) \ldots \mathcal{O}(x_n) \rangle. \tag{3.41}$$

In the CFT this is done via the formula

$$\langle \mathcal{O}(x_1) \ldots \mathcal{O}(x_n) \rangle = \frac{\delta^n \log Z_{\text{CFT}}[J]}{\delta J(x_1) \ldots \delta J(x_n)} \bigg|_{J=0}, \tag{3.42}$$

where

$$Z_{\text{CFT}}[J] = \left\langle \exp\left(\int J(x)\mathcal{O}(x) \right) \right\rangle. \tag{3.43}$$

The AdS/CFT correspondence states then that the CFT generating functional with source $J = \phi_0 = \phi(0, x)$ is equal to the path integral on the gravity side evaluated over a bulk field which has the value ϕ_0 at the boundary of AdS [35, 36]. We write

$$Z_{\text{CFT}}[\phi_0] = \left\langle \exp\left(\int \phi_0(x)\mathcal{O}(x) \right) \right\rangle = \int_{\phi \to \phi_0} \mathcal{D}\phi \exp(S_{\text{grav}}). \tag{3.44}$$

In the limit in which classical gravity is a good approximation we use the classical on-shell gravity action, i.e.

$$Z_{CFT}[\phi_0] = \exp(S_{grav}^{on-shell}[\phi \longrightarrow \phi_0]). \tag{3.45}$$

After holographic renormalization the on-shell action is renormalized and the AdS/CFT prescription becomes

$$Z_{CFT}[\phi_0] = \exp(S_{grav}^{renor}[\phi \longrightarrow \phi_0]). \tag{3.46}$$

The correlation functions are then renormalized as

$$\langle \mathcal{O}(x_1)...\mathcal{O}(x_n) \rangle = \frac{\delta^n S_{grav}^{renor}[\phi \longrightarrow \phi_0]}{\delta \varphi(x_1)...\delta \varphi(x_n)}\Big|_{\varphi=0}. \tag{3.47}$$

See for example [51].

3.7 The geometry/entanglement connection

3.7.1 Ryu–Takayanagi formula

The Ryu–Takayanagi formula [26, 27] is a generalization of the Bekenstein–Hawking formula, based on the AdS/CFT correspondence, in which we identify the entanglement entropy in $(d + 1)$-dimensional QFT with a geometric quantity in $(d + 2)$-dimensional gravity.

We consider the metric in AdS_{d+2} given by

$$ds^2 = g^{\mu\nu}dx_\mu dx_\nu = \frac{R^2}{z^2}\left(-dt^2 + \sum_{i=1}^{d} dx_i^2 + dz^2\right). \tag{3.48}$$

The dual $(d + 1)$-dimensional CFT lives on the boundary located at $z = 0$. The radial coordinate z is a lattice spacing and the theory on the boundary should be properly understood as the continuum limit (in the sense of RG) of a regularize CFT, i.e. with a cutoff Λ. This regularized CFT lives on a surface $z = a$, where $a = 1/\Lambda$ and $a \longrightarrow 0$.

Our observers live on the boundary $z = a \longrightarrow 0$ of AdS_{d+2}. The accessible region A and the inaccessible region B are both on this boundary $z = a$.

The entanglement entropy in the CFT_{d+1} which lives on this boundary can be computed from the gravity theory which lives in the bulk AdS_{d+2} by the formula

$$S_A = \frac{\text{Area}(\Gamma_A)}{4G_N^{(d+2)}}. \tag{3.49}$$

The Γ_A is a minimal area surface in the time slice M which is the extension of $\mathbf{N} = A \cup B$ with boundary $\partial \Gamma_A = \partial A$. See figure 3.4.

As an example we consider AdS_3/CFT_2. In this case we are interested in the line segment $x \in [-l/2, l/2]$ on $z = a$. This segment is extended in the bulk to a circle (which we can check is the line of minimal length) given by

$$z = \frac{l}{2}\sin s, \quad x = \frac{l}{2}\cos s, \quad \epsilon \leqslant s \leqslant \pi - \epsilon, \quad \epsilon = \frac{2a}{l} \longrightarrow 0. \tag{3.50}$$

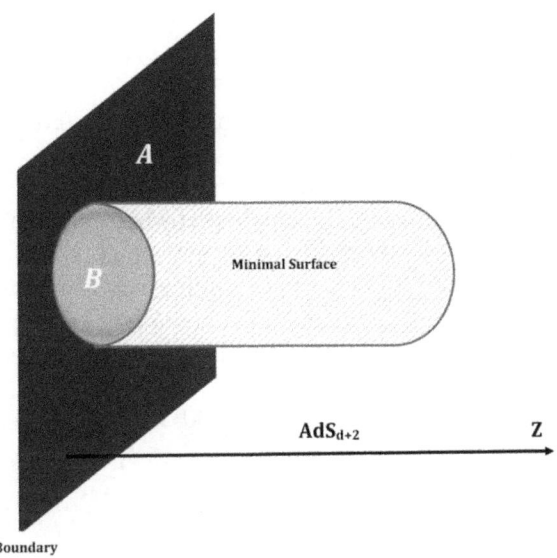

Figure 3.4. The accessible and inaccessible regions in AdS.

The length of the circle is given by $L = 2R \ln l/a$. The entanglement entropy becomes

$$S_A = \frac{L}{4G_N^{(3)}} = \frac{c}{3} \ln \frac{l}{a}. \qquad (3.51)$$

The central charge c of CFT_2 is related to the radius R of AdS_3 by the relation [52]

$$c = \frac{3R}{2G_N^{(3)}}. \qquad (3.52)$$

The calculation on the field theory side is more involved. For a conformal field theory in two dimensions with central charge c on the torus we find explicitly [53]

$$S_A = \frac{c}{3} \ln \frac{\Sigma}{\epsilon}. \qquad (3.53)$$

3.7.2 The CFT/BH correspondence

The CFT/BH correspondence between conformal field theories and black holes is the most understood case of the holographic gauge/gravity duality.

Let $|\psi(0)\rangle$ be the vacuum state of the CFT_d. This state is dual to pure AdS metric given by (Poincaré coordinates)

$$ds^2 = \frac{L^2}{z^2}(dz^2 + dx_\mu dx^\mu). \qquad (3.54)$$

Let $|\psi(\zeta)\rangle$ be a one-parameter family of CFT_d excited states which are dual to the perturbed metrics (Fefferman–Graham coordinates)

$$ds^2 = \frac{L^2}{z^2}(dz^2 + dx_\mu dx^\mu + z^d \bar{h}_{\mu\nu}(z,x)dx^\mu dx^\nu). \tag{3.55}$$

This corresponds to a spacetime M_ζ with boundary at $z \to 0$ denoted by ∂M_ζ where the state $|\psi(\zeta)\rangle$ is living.

This metric can also be understood as corresponding to a spacetime M_ζ, which is a perturbation of pure AdS, dual to a small perturbation $|\psi(\zeta)\rangle_{\zeta \to 0}$ of the CFT$_d$ vacuum $|\psi(0)\rangle$. This is an asymptotically AdS spacetime.

For higher excited states $|\psi(\zeta)\rangle$ we cannot assume a classical supergravity solution since l_s is no longer much less than L and as a consequence stringy corrections of the order l_s^2 and higher become important. The geometry (and even the topology) of AdS$_{d+1}$ becomes therefore very different.

As an example we consider the Schwarzschild–AdS black hole in $d+1$ dimensions given by the metric

$$ds^2 = -f_M dt^2 + \frac{dr^2}{f_M} + r^2 d\Omega_{d-2}, \quad f_M = 1 - \frac{2\mu}{r^{d-3}} + \frac{r^2}{L^2}. \tag{3.56}$$

The Penrose diagram of the Schwarzschild–AdS spacetime is shown in figure 3.5. The two light-like infinities $\mathcal{J}^\pm$ and the space-like infinity i^0 are replaced with the universal covering of global AdS spacetime in both regions I and III. The asymptotic regions are denoted by A and B and they are the conformal boundary of AdS, i.e. $\mathbf{R} \times \mathbf{S}^{d-1}$. The regions I and III are causally disconnected but signals from region I can intersect with signals from region III behind the horizon. This is then a two-sided black hole, i.e. with two different exterior regions, which can also be viewed as a wormhole.

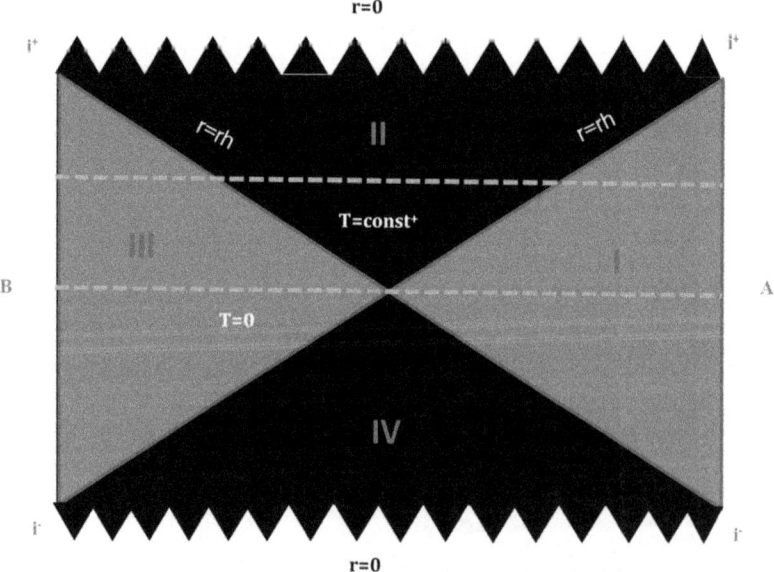

Figure 3.5. The Schwarzschild–AdS spacetime.

The Schwarzschild–AdS black hole is conjectured to be dual to the thermofield double state $|\Psi\rangle$ given by [23]

$$|\Psi\rangle = \frac{1}{\sqrt{Z(\beta)}} \sum_i \exp(-\beta E_i/2)|E_i^A\rangle \otimes |E_i^B\rangle. \tag{3.57}$$

This is the vacuum state of two identical non-interacting copies of the conformal field theory living on the boundaries A and B (which can only interact via entanglement).

The Schwarzschild–AdS black hole is a connected spacetime in which light signals traveling from A and B can intersect. This connectedness is equivalent to the entanglement between the two subsystems Q_A and Q_B, which contain the degrees of freedom of the local CFT living on A and B, as expressed by the thermofield double state $|\Psi\rangle$.

The reduced density matrix for the subsystem Q_A is immediately given by integrating out the degrees of freedom associated with the subsystem Q_B as $\rho_A = tr_B|\Psi\rangle\langle\Psi|$ with a non-zero entanglement entropy $S_A = -tr_A \rho_A \log \rho_A$.

In the limit of zero temperature the reduced density matrix ρ_A approaches $\rho_A = |E_0^A\rangle\langle E_0^A|$ and as a consequence the entanglement entropy vanishes, i.e.

$$|\Psi\rangle \longrightarrow |\Phi\rangle = |E_0^A\rangle \otimes |E_0^B\rangle, \quad T \longrightarrow 0. \tag{3.58}$$

This is dual to a spacetime consisting of the product of two pieces corresponding to the two disconnected regions A and B, i.e. light signals traveling from A and B cannot intersect. Thus in the limit $T \longrightarrow 0$ entanglement between Q_A and Q_B is removed and correspondingly the connectivity between regions A and B is broken. This shows clearly that entanglement between Q_A and Q_B is a necessary condition for classical connectivity between A and B [21].

3.7.3 The first law of entanglement and Einstein's equations

We consider a one-parameter family of CFT states $|\psi(\zeta)\rangle$. The dual spacetime of the vacuum state $|\psi(0)\rangle$ is pure AdS. The dual spacetime $\mathcal{M}_\psi$ of the perturbed state $|\psi(\zeta)\rangle$ is a perturbation of AdS with boundary $\partial\mathcal{M}_\psi$, which is a perturbation of Minkowski spacetime.

The accessible region B is a ball shaped region of radius R on the boundary $\partial\mathcal{M}_\psi$.

The entanglement entropy of B in the CFT is equal to the von Neumann entropy of the reduced density matrix $\rho_B = tr_{\bar{B}}|\psi(\zeta)\rangle\langle\psi(\zeta)|$. We compute immediately

$$\frac{d}{d\zeta}S_B = \frac{d}{d\zeta}E_B^{\text{Hyp}}. \tag{3.59}$$

The expectation value of the modular Hamiltonian $H_B = -\log \rho_B|_{\zeta=0}$ is the hyperbolic energy E_B^{Hyp}.

This is effectively the first law of thermodynamics $dE = dS$ which holds in the CFT for arbitrary perturbations ζ of the vacuum state $|\psi(0)\rangle$.

The holographic extension of this law to the bulk gives (after a very long calculation) precisely Einstein's equation around AdS background. This is the prime example of how spacetime geometry is equivalent to quantum entanglement [28–31].

3.8 Emergent time from entanglement verified experimentally

The quantum measurement and the associated irreversible and instantaneous collapse of the wave function seem to lie at the heart of the problem of time. The only other fundamental principle in nature which is known to be irreversible is the second law of thermodynamics which states that the entropy of a closed system can only increase over time. The entropy of the Universe is thus constantly increasing with the cosmological expansion from its very small starting value at the initial big bang singularity. The smallness of the initial entropy is in itself quite a mysterious property [54]. On the other hand, the constant increase of entropy with the expansion of the Universe is thought to be correlated with the observed arrow of time which is directed from past to future in the same way that the quantum measurement as well as causality are directed from past to future.

Thus, the problem of time is reduced to the problem of the flow of time, which in turn is reduced to the problem of the arrow of time, and then reduced even further to the law of the increase of entropy. It seems that this can be reduced even further to quantum entanglement, which is incidentally measured by entropy. Measurement involves the entanglement between the observer and the observed in an essential way.

The problem of time [55–58], in its simplest form, is the realization that the Wheeler–De Witt equation [59] describes a static universe without any time. This is because in the canonical formulation of general relativity the Hamiltonian is a Dirac constraint. Thus in this almost unique instance, quantum mechanics (with its absolute time) and general relativity (with its dynamical time) are merged successfully (the Wheeler–De Witt is the Schrödinger equation for the metric without any UV divergences), only at the cost of dispensing altogether with the notion of time.

However, time must emerge again somehow from this equation in order to make contact with the observed time evolution of the Universe. A solution of this problem is given by Page and Wootters [60] (see also [61]) in which time emerges dynamically from quantum entanglement. This goes as follows. The universe as a whole is a static system S given by a static state vector $|\Psi\rangle$ as dictated by the Wheeler–De Witt equation, namely $H|\Psi\rangle = 0$, where H is the Hamiltonian of the Universe. This static system S is divided into two entangled subsystems: a subsystem C which is called a clock, and the rest of the system given by the subsystem R. We will assume that we can neglect the interactions between these subsystems and hence the Hamiltonian can be rewritten as

$$H = H_C \otimes 1_R + 1_R \otimes H_R. \tag{3.60}$$

The external observer sees no time since his Schrödinger equation is $H|\Psi\rangle = 0$. The subsystems C and R both obey Schrödinger's equation with respect to an internal observer whose time parameter is given by the clock C. The state vector of the clock may be written as

$$|\Phi(t)\rangle_c = \exp(-iH_C t/\hbar)|\Phi(0)\rangle_C. \qquad (3.61)$$

Thus, we can immediately compute (since $H = 0$ and $[H_R, H_C] = 0$)

$$|\psi(t)\rangle_R = \langle\Phi_C(t)|\Psi\rangle = \exp(-iH_R t/\hbar)\langle\Phi_C(0)|\Psi\rangle = \exp(-iH_R t/\hbar)|\psi(0)\rangle_R. \qquad (3.62)$$

In other words, the subsystem R is seen by the internal observer, using the clock C, as evolving in time t. Hence, quantum entanglement between the chosen clock and the rest of the Universe yields a static universe with respect to a hypothetical external observer, and an evolving universe with respect to an internal observer (perhaps the clock itself). The internal observer detects the evolution by inspecting the quantum correlations between the clock C and the subsystem R, whereas the external observer inspects the global properties of the entangled state of C and R. It is clearly seen that what is called here a clock is what Wigner calls a conscious mind entangled with, and performing measurements on, the rest of the world.

The Page–Wootters solution of the problem of time is very much in the spirit of the relational interpretation of quantum mechanics [62] which all fall, on the surface of it, within the philosophical relationalism of space and time advocated by Leibniz against Newton's absolutism and Kant's transcendental idealism. Thus, on this view, time is not absolute (Newton) nor it is an intuition of the cognition (Kant) but it is a relation to clocks. But as we stated above clocks are intertwined inextricably with internal observers as in all interpretations of quantum mechanics.

This elegant and simple scenario for the emergence of time from quantum entanglement was verified experimentally in the truly beautiful experiment of [63]. In this experiment the Universe is constituted of only two entangled photons prepared in the initial state

$$|\Psi\rangle = \frac{1}{\sqrt{2}}(|H\rangle_C |V\rangle_R - |V\rangle_C |H\rangle_R). \qquad (3.63)$$

This satisfies the Wheeler–De Witt equation with $H_C = H_R = i\hbar\omega(|H\rangle\langle V| - |V\rangle\langle H|)$, where ω is the timescale of the model. Thus, if the clock photon is in a horizontal polarization state, the other photon will be in a vertical polarization state, and vice versa. The two photons are sent along two separate paths through identical birefringent plates which cause rotations of the polarization of the photons. The amount of rotation incurred is proportional to the time spent by the photon in the plates. Thus, the thickness of the plates is a measure of the coordinate time.

The clock photon has only two time readings: t_1 corresponding to a horizontal polarization (detector 1 clicking) and t_2 corresponding to a vertical polarization (detector 2 clicking). The internal observer who measures the clock photon in detectors 1 and 2 also measures the other photon's polarization in detectors 3 (vertical) and 4 (horizontal). These measurements are expressed by the conditional probability $P_{3|x} = p(t_x)$: the probability that the other photon has vertical polarization (detector 3 clicking) if the clock detector x clicks. This probability is given exactly by the average number of coincidences between the detectors 3 and x over the total number of photon pairs from all detectors.

The internal observer can only access the clock time t_x and not the coordinate time. This is achieved experimentally by changing in a random way the thickness of the birefringent plates in the path of the clock photon, repeating the same conditional probability measurement, and then taking the average over the thickness. This allows us to obtain the probabilities $P_{3|x}^{\tau_i} = p(t_x + \tau_i)$, with $\tau_i = \delta_i/\omega_i$, where δ_i is the optical thickness. In this so-called 'observer mode', it is seen that the internal observer becomes entangled with the photon clock, and thus she can see an evolution of the other photon's state as a function of the clock time.

The experimental result of the average number of coincidences as a function of time is found to be in excellent agreement with the theory [63] which, for a global state ρ, is given by the conditional probability

$$P(d, t) = \frac{\int dT Tr P_{d,t}(T)\rho}{\int dT P_t(T)\rho}. \tag{3.64}$$

$P_t(T)$ is the projector corresponding to a measurement of a clock time t at the coordinate instant T, and $P_{d,t}(T)$ is the projector corresponding to the measurement t of the clock photon, and to measurement d of the other photon, at the coordinate instant T. The fact that we integrate over T means that this conditional probability does not depend on the coordinate time.

On the other hand, in the so-called 'super-observer mode', it is shown that the global entangled state of the two photons remains the same for any thickness of the birefringent plates which represents, as we said, the external coordinate time. Thus, although the external observer has access to the thickness of the plates (coordinate time), she is required to coherently quantum erase [64, 65] the which-way information, i.e. the clock time measurements of the internal observer obtained in the measurement of the clock photon polarization, in order to avoid becoming herself entangled with the clock photon. The global state is found to be indeed static with respect to coordinate time.

3.9 Other models of quantum gravity

3.9.1 Causal dynamical triangulation versus Horava–Lifshitz gravity[1]

Another powerful approach to the emergence of time and spacetime is Lorentzian causal dynamical triangulation [67–69]. In this approach spacetime is built out of four-simplices (generalization of two-simplices, i.e. triangles, to four dimensions) which are equipped with a flat Minkowski metric. The causality requirement singles out globally hyperbolic manifolds which admit a global proper-time foliation structure and as a consequence Wick rotation to Euclidean is meaningful. The Hilbert–Einstein action is given in this discrete setting by the Regge action [70]. The path integral is obtained as the sum over the set of all causal triangulations weighted with the Regge action. The parameters of the model are Newton's gravitational

---
[1] This section is taken verbatim from [66].

constant G and the cosmological constant Λ which appear as the parameters K_0 and K_4 in the Regge action. Also the model depends on two more parameters given by the lengths of time-like and spatial-like links a_t and a_s, respectively. We have $a_t^2 = \alpha a_s^2$, where the asymmetry factor $\alpha < 0$ appears as a parameter Δ in the Regge action.

Causal dynamical triangulation (CDT) is intimately related to Horava–Lifshitz (HL) gravity [71–73], which also, like CDT, assumes global time foliation and introduces anisotropy between space and time, but in such a way as to achieve power-counting renormalizability of quantum gravity. This theory is effectively a generalization to gravity of the d-dimensional Lifshitz scalar field theory given by the Lifshitz–Landau free energy density [74]

$$S = a_2\phi^2 + a_4\phi^4 + \cdots + c_2(\partial_\alpha\phi)^2 + d_2(\partial_\beta\phi)^2 + e_2(\partial_\beta^2\phi)^2 + \cdots \qquad (3.65)$$

The anisotropy is introduced by the distinction between the indices $\beta = 1, \ldots, m$ and $\alpha = m + 1, \ldots, d$. The three phases present in this theory are: helicoidal ($|\partial_t\phi(x)| < 0$), paramagnetic ($\phi(x) = 0$) and ferromagnetic ($|\phi(x)| > 0$). The phase diagram is depicted in figure 3.6(a).

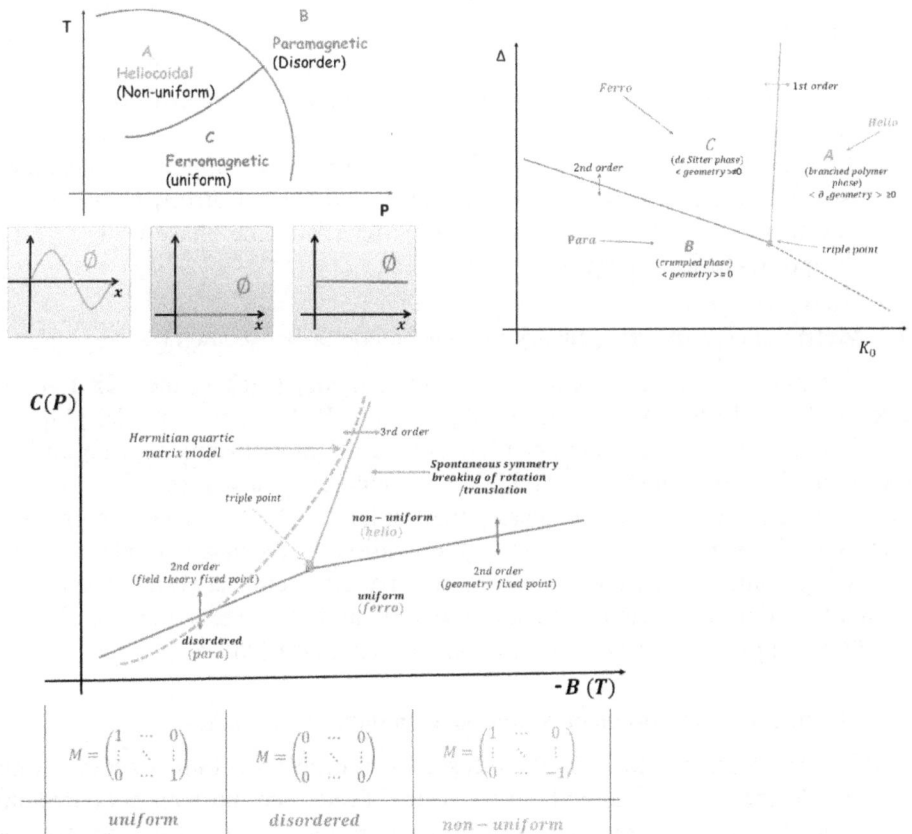

Figure 3.6. The phase diagrams of causal dynamical triangulation [75, 76], Lifshitz scalar field theory [74] and multitrace matrix model [66], copyright (2017) World Scientific.

The phase structure of causal dynamical triangulation is also summarized in figure 3.6(b). The cosmological constant K_4 which controls the total volume is fixed at its critical value and the phase diagram is then drawn in the plane $K_0 - \Delta$, where K_0 is proportional to the inverse bare gravitational coupling constant G while Δ is effectively the asymmetry factor α. There are three distinct phases [75, 76]:

- The de Sitter spacetime phase C. This is the analogue of the ferromagnetic phase ($d_2 > 0$, $a_2 < 0$) of the Lifshitz scalar field theory or the ordered phase in noncommutative scalar field theory (see below).
- The crumpled phase B where neither space nor time have any extent and therefore there is no geometry. This is the analogue of the paramagnetic phase ($d_2 > 0$, $a_2 > 0$) of the Lifshitz scalar field theory or the disordered phase in noncommutative scalar field theory (see below).
- The branched polymer phase A where the geometry oscillates in time. This is the analogue of the helicoidal phase ($d_2 < 0$) in Lifshitz scalar field theory or the non-uniform ordered phase in noncommutative scalar field theory (see below).
- The transition from A to C is first order whereas the transition from B to C could be either first order or second order and as a consequence there is a possibility of a continuum limit. Similarly, the transition between the ferromagnetic (C) and paramagnetic (B) phases in the Lifshitz scalar field theory although usually second order it could be first order. As we will see a strikingly similar situation occurs in noncommutative scalar field theory (see below).
- Also in both theories CDT and HL the spectral dimension at short distances is 2 and only becomes 4 at large distances and the anisotropy between space and time disappear in CDT in the de Sitter spacetime phase while in HL it disappears at low energies.

3.9.2 Matrix models of string theory and noncommutative geometry

The combination of the matrix models of string theory (such as the IKKT matrix model [48], the BFSS matrix model [47] and the BMN matrix model [49]) and Conne's noncommutative geometry [77] provide the most promising (the author's view) candidate of quantum gravity as an emerging structure. The quantum entanglement of quantum mechanics is due fundamentally to the noncommutative structure of the phase space after Dirac quantization. In the matrix model approach the starting point is precisely a noncommutative structure between the spacetime coordinates. For a systematic review see [79] and references therein. A good representation of this approach is the work of Steinacker [78].

3.9.3 The measurement problem within noncommutative geometry

It will be very interesting to study the measurement problem within noncommutative geometry. A starting point could be the viewing of quantum mechanics (noncommutative geometry) as an approximation to classical mechanics (classical geometry). See for example [80]. In this approach the measurement problem seems to be intimately connected to the UV–IR mixing of noncommutative field theory [81].

References

[1] Birkhoff G D and Langer R 1923 *Relativity and Modern Physics* (Cambridge, MA: Harvard University Press)
Jebsen J T 1921 *Ark. Mat. Astron. Fys.* **15** 1
Alexandrow W 1923 *Ann. Physik* **72** 141
[2] Unruh W G 1976 Notes on black hole evaporation *Phys. Rev.* D **14** 870
[3] Susskind L and Lindesay J 2005 *An Introduction to Black Holes, Information and the String Theory Revolution: The Holographic Universe* (Hackensack, NJ: World Scientific) p 183
[4] Mukhanov V and Winitzki S 2007 *Introduction to Quantum Effects in Gravity* (Cambridge: Cambridge University Press)
[5] Bekenstein J D 1973 Black holes and entropy *Phys. Rev.* D **7** 2333
[6] Bekenstein J D 1974 Generalized second law of thermodynamics in black hole physics *Phys. Rev.* D **9** 3292
[7] Jacobson T 2003 Introduction to quantum fields in curved space-time and the Hawking effect arXiv: gr-qc/0308048
[8] Hawking S W 1975 Particle creation by black holes *Commun. Math. Phys.* **43** 199
Hawking S W 1976 Particle creation by black holes *Commun. Math. Phys.* **46** 206
[9] Hawking S W 1976 Breakdown of predictability in gravitational collapse *Phys. Rev.* D **14** 2460
[10] Page D N 1993 Information in black hole radiation *Phys. Rev. Lett.* **71** 3743
[11] Page D N 2013 Time dependence of Hawking radiation entropy *J. Cosmol. Astropart. Phys.* **1309** 028
[12] 't Hooft G 1985 On the quantum structure of a black hole *Nucl. Phys.* B **256** 727
[13] 't Hooft G 1990 The black hole interpretation of string theory *Nucl. Phys.* B **335** 138
[14] Susskind L, Thorlacius L and Uglum J 1993 The stretched horizon and black hole complementarity *Phys. Rev.* D **48** 3743
[15] Almheiri A, Marolf D, Polchinski J and Sully J 2013 Black holes: complementarity or firewalls? *J. High Energy Phys.* **1302** 062
[16] Maldacena J and Susskind L 2013 Cool horizons for entangled black holes *Fortsch. Phys.* **61** 781
[17] Susskind L 2016 Copenhagen vs Everett, teleportation, and ER=EPR *Fortsch. Phys.* **64** 551
[18] Susskind L 2016 ER=EPR, GHZ, and the consistency of quantum measurements *Fortsch. Phys.* **64** 72
[19] Einstein A and Rosen N 1935 The particle problem in the general theory of relativity *Phys. Rev.* **48** 73
[20] Einstein A, Podolsky B and Rosen N 1935 Can quantum mechanical description of physical reality be considered complete? *Phys. Rev.* **47** 777
[21] Van Raamsdonk M 2010 Building up spacetime with quantum entanglement *Gen. Rel. Grav.* **42** 2323
Van Raamsdonk M 2010 Building up spacetime with quantum entanglement *Int. J. Mod. Phys.* D **19** 2429
[22] Israel W 1976 Thermo field dynamics of black holes *Phys. Lett.* A **57** 107
[23] Maldacena J M 2003 Eternal black holes in anti-de Sitter *J. High Energy Phys.* **0304** 021
[24] Hartle J B and Hawking S W 1976 Path integral derivation of black hole radiance *Phys. Rev.* D **13** 2188
[25] Maldacena J M 1999 The large N limit of superconformal field theories and supergravity *Int. J. Theor. Phys.* **38** 1113

Maldacena J M 1998 The large N limit of superconformal field theories and supergravity *Adv. Theor. Math. Phys.* **2** 231

[26] Ryu S and Takayanagi T 2006 Holographic derivation of entanglement entropy from AdS/CFT *Phys. Rev. Lett.* **96** 181602
[27] Ryu S and Takayanagi T 2006 Aspects of holographic entanglement entropy *J. High Energy Phys.* **0608** 045
[28] Lashkari N, McDermott M B and Van Raamsdonk M 2014 Gravitational dynamics from entanglement 'thermodynamics' *J. High Energy Phys.* **1404** 195
[29] Faulkner T, Guica M, Hartman T, Myers R C and Van Raamsdonk M 2014 Gravitation from entanglement in holographic CFTs *J. High Energy Phys.* **1403** 051
[30] Van Raamsdonk M 2016 Lectures on gravity and entanglement arXiv:1609.00026 [hep-th]
[31] Casini H, Huerta M and Myers R C 2011 Towards a derivation of holographic entanglement entropy *J. High Energy Phys.* **1105** 036
[32] t Hooft G 1993 Dimensional reduction in quantum gravity *Conf. Proc.* C **930308** 284
[33] Susskind L 1995 The World as a hologram *J. Math. Phys.* **36** 6377
[34] Bousso R 2002 The holographic principle *Rev. Mod. Phys.* **74** 825
[35] Gubser S S, Klebanov I R and Polyakov A M 1998 Gauge theory correlators from noncritical string theory *Phys. Lett.* B **428** 105
[36] Witten E 1998 Anti-de Sitter space and holography *Adv. Theor. Math. Phys.* **2** 253
[37] 't Hooft G 1974 A planar diagram theory for strong interactions *Nucl. Phys.* B **72** 461
[38] Polchinski J 1995 Dirichlet Branes and Ramond–Ramond charges *Phys. Rev. Lett.* **75** 4724
[39] Gibbons G W and Maeda K 1988 Black holes and membranes in higher dimensional theories with dilaton fields *Nucl. Phys.* B **298** 741
[40] Itzhaki N, Maldacena J M, Sonnenschein J and Yankielowicz S 1998 Supergravity and the large N limit of theories with sixteen supercharges *Phys. Rev.* D **58** 046004
[41] Horowitz G T and Strominger A 1991 Black strings and P-branes *Nucl. Phys.* B **360** 197
[42] Horowitz G T and Polchinski J 2006 Gauge/gravity duality *Approaches to Quantum Gravity* ed D Oriti (Cambridge: Cambridge University Press) pp 169–86
[43] Wilson K G 1974 Confinement of quarks *Phys. Rev.* D **10** 2445
[44] Aharony O, Bergman O, Jafferis D L and Maldacena J 2008 $N = 6$ superconformal Chern–Simons-matter theories, M2-branes and their gravity duals *J. High Energy Phys.* **0810** 091
[45] Dijkgraaf R, Verlinde E P and Verlinde H L 1997 Matrix string theory *Nucl. Phys.* B **500** 43
[46] Gregory R and Laflamme R 1993 Black strings and p-branes are unstable *Phys. Rev. Lett.* **70** 2837
[47] Banks T, Fischler W, Shenker S H and Susskind L 1997 M theory as a matrix model: a conjecture *Phys. Rev.* D **55** 5112
[48] Ishibashi N, Kawai H, Kitazawa Y and Tsuchiya A 1997 A large N reduced model as superstring *Nucl. Phys.* B **498** 467
[49] Berenstein D E, Maldacena J M and Nastase H S 2002 Strings in flat space and pp waves from $N = 4$ super Yang–Mills *J. High Energy Phys.* **0204** 013
[50] Hanada M 2016 What lattice theorists can do for superstring/M-theory *Int. J. Mod. Phys.* A **31** 1643006
[51] Ramallo A V 2015 Introduction to the AdS/CFT correspondence *Lectures on Particle Physics, Astrophysics and Cosmology* Springer Proceedings in Physics (Berlin: Springer) pp 411–74
[52] Brown J D and Henneaux M 1986 Central charges in the canonical realization of asymptotic symmetries: an example from three-dimensional gravity *Commun. Math. Phys.* **104** 207

[53] Holzhey C, Larsen F and Wilczek F 1994 Geometric and renormalized entropy in conformal field theory *Nucl. Phys.* B **424** 443
[54] Penrose R 1989 *The Emperor's New Mind* (Oxford: Oxford University Press)
Penrose R 2004 *The Road to Reality* (London: Jonathan Cape)
[55] Isham C J 1992 Canonical quantum gravity and the problem of time arXiv: gr-qc/9210011
[56] Sorkin R D 1994 On the role of time in the sum over histories framework for gravity *Int. J. Theor. Phys.* **33** 523
[57] Ashtekar A 2005 Gravity and the quantum *New J. Phys.* **7** 198
[58] Anderson E 2010 The problem of time in quantum gravity arXiv:1009.2157 [gr-qc]
[59] DeWitt B S 1967 Quantum theory of gravity. 1. The canonical theory *Phys. Rev.* **160** 1113
[60] Page D N and Wootters W K 1983 Evolution without evolution: dynamics described by stationary observables *Phys. Rev.* D **27** 2885
[61] Gambini R, Porto R A, Pullin J and Torterolo S 2009 Conditional probabilities with Dirac observables and the problem of time in quantum gravity *Phys. Rev.* D **79** 041501
[62] Rovelli C 1996 Relational quantum mechanics *Int. J. Theor. Phys.* **35** 1637–78
[63] Moreva E, Brida G, Gramegna M, Giovannetti V, Maccone L and Genovese M 2014 Time from quantum entanglement: an experimental illustration *Phys. Rev.* A **89** 052122
[64] Scully M O, Englert B G and Walther H 1991 Quantum optical tests of complementarity *Nature* **351** 111–6
[65] Durr S, Nonn T and Rempe G 1998 Origin of quantum-mechanical complementarity probed by a 'which-way' experiment in an atom interferometer *Nature* **395** 33–7
[66] Ydri B, Soudani C and Rouag A 2017 Quantum gravity as a multitrace matrix model *Int. J. Mod. Phys.* A **32** 07724
[67] Ambjorn J, Jurkiewicz J and Loll R 2005 Reconstructing the Universe *Phys. Rev.* D **72** 064014
[68] Ambjørn J, Janik R, Westra W and Zohren S 2006 The emergence of background geometry from quantum fluctuations *Phys. Lett.* B **641** 94
[69] Ambjorn J, Gorlich A, Jurkiewicz J and Loll R 2008 Planckian birth of the quantum de Sitter universe *Phys. Rev. Lett.* **100** 091304
[70] Regge T 1961 General relativity without coordinates *Nuovo Cim.* **19** 558
[71] Horava P 2009 Spectral dimension of the Universe in quantum gravity at a Lifshitz point *Phys. Rev. Lett.* **102** 161301
[72] Horava P 2009 Membranes at quantum criticality *J. High Energy Phys.* **0903** 020
[73] Horava P 2009 Quantum gravity at a Lifshitz point *Phys. Rev.* D **79** 084008
[74] Goldenfeld N 1992 *Lectures on Phase Transitions and the Renormalization Group* Frontiers in Physics vol 85 (Reading, MA: Addison-Wesley) p 394
[75] Ambjorn J, Gorlich A, Jordan S, Jurkiewicz J and Loll R 2010 CDT meets Horava–Lifshitz gravity *Phys. Lett.* B **690** 413
[76] Gorlich A 2011 Causal dynamical triangulations in four dimensions arXiv:1111.6938
[77] Connes A 1996 Gravity coupled with matter and foundation of noncommutative geometry *Commun. Math. Phys.* **182** 155
[78] Steinacker H 2011 Non-commutative geometry and matrix models *PoS QGQGS* **2011** 004
[79] Ydri B 2018 *Matrix Models of String Theory* (Bristol: IOP Publishing)
[80] Bracken A J 2003 Quantum mechanics as an approximation to classical mechanics in Hilbert space *J. Phys.* A **36** L329–35
[81] Minwalla S, Van Raamsdonk M and Seiberg N 2000 Noncommutative perturbative dynamics *J. High Energy Phys.* **02** 020

IOP Publishing

Philosophy and the Interpretation of Quantum Physics

Badis Ydri

Chapter 4

Quantum dualism

Starting in this chapter we provide a synthesis based on the quantum dualism, i.e. the fact that the Copenhagen interpretation provides the local view of Reality whereas the many-worlds formalism provides the manifold view and the two views are complementary not contradictory.

Thus in this chapter we put forward our first synthesis in which the quantum dualism is emphasized over unitarity. In fact it is believed that the quantum dualism itself is unitary via the Copenhagen/many-worlds correspondence, i.e. the complementarity between the local view of Reality provided by the Copenhagen first-person observers and the manifold view of Reality provided by the many-worlds third-person observers.

In the synthesis which will follow in the next chapter we will take the inverse approach and stress unitarity more explicitly than quantum dualism, whereas in the final chapter of this book we will return to quantum dualism and relate it to perspectivism and other implicit and explicit characteristics of the Copenhagen interpretation.

In this chapter we also draw systematic parallels between the measurement problem in quantum mechanics and the information loss problem in black holes. Then we proceed to propose a solution of the former along the lines of the solution of the latter which is based on the holographic gauge/gravity duality. The proposed solution is based on (i) the quantum dualism and (ii) the properties of quantum entanglement, in particular its fungibility. This topic is further discussed in the next chapter.

4.1 The measurement problem as an information loss problem

The reduction/collapse of the state vector is formally very similar to the information loss problem. Indeed, they both involve going from a pure state to a mixed state which results in a loss of quantum entanglement and a decrease of information. In summary, we have

- *The measurement problem*: $\rho_e \longrightarrow \rho_r$ (inaccessible degrees of freedom of the environment).
- *The information loss problem*: $\rho_{in} \longrightarrow \longrightarrow \rho_{out}$ (inaccessible degrees of freedom of the information behind the horizon).

In more detail we have

$$\rho_e = |\Phi_e\rangle\langle\Phi_e| \longrightarrow \rho_r = \rho_{S+D} = \mathrm{Tr}_E|\Psi\rangle\langle\Psi|, \quad |\Psi\rangle = U|\Phi_e\rangle|E_0\rangle. \qquad (4.1)$$

$$\rho_{in} = |\psi_{in}\rangle\langle\psi_{in}| \longrightarrow \rho_{out} = \rho_{\mathrm{Hawking}} = \mathrm{Tr}_S|\psi_{out}\rangle\langle\psi_{out}|, \quad |\psi_{out}\rangle = S|\psi_{in}\rangle. \qquad (4.2)$$

The environment E acts therefore as an observer (Wigner for example) performing a measurement on the joint system $S + D$, where S is the physical system under observation (Schrödinger's cat, for example) and D is the detector which also acts as an observer (Wigner's friend). The joint system $S + D + E$ evolves by means of a unitary matrix U since the three parts S, D and E are highly entangled.

The environment E contains therefore the inaccessible degrees of freedom of the system which lie at the root of the collapse of the wave function in the same way that the inaccessible degrees of freedom which went behind the horizon lie at the root of the information loss problem. In other words, the environment E plays the role of the interior of the black hole whereas the joint system $S + D$ plays the role of the exterior of the black hole with an effective horizon separating them. In fact the measurement process (like a radiating black hole) can always be thought of as involving two entangled systems which are separated with a horizon where one of the systems (the observed system) contains the accessible degrees of freedom whereas the other one (the observing system) contains the inaccessible degrees of freedom which, since they are inaccessible, are required to be traced out yielding as a consequence a sort of information loss or collapse of the state vector.

Of course the environment in this argument can be replaced with consciousness (for those who think that the observer's mind is what precipitates collapse [1, 2]) or quantum gravity (for those who think that spacetime curvature at the Planck scale is what precipitates an objective collapse [3]).

By the holographic gauge/gravity duality [4] the black hole is in fact dual to a certain unitary conformal field theory (CFT) and hence the information loss problem is only a coarse-graining effect since the information will eventually start to emerge with the radiation at the Page time when the entanglement entropy reaches its maximum (Page curve). A very explicit example is the case of a Schwarzschild–AdS black hole which is conjectured to be dual to the thermofield double state $|\Psi\rangle$ of a pair of CFTs living on the conformal boundaries of AdS space [5] (see also [6]).

A rigorous confirmation, using non-perturbative Monte Carlo methods, of the fact that black holes follow the predictions of the AdS/CFT correspondence is carried out in [7]. This is the case of M-(atrix) theory, i.e. the case of the celebrated BFSS matrix model [8] which defines the dual gauge theory of a black 0-brane as given by a bound state of N D0-branes. In this case it is explicitly seen that in the limit $N \longrightarrow \infty$ both the Hawking radiation (as computed from the matrix model on the gauge side) and the measurement problem (since the theory becomes classical supergravity on the gravity side) disappear.

4.2 The measurement problem and quantum entanglement

The measurement problem, similarly to the information loss problem, is not considered here to be a fundamental problem but only a coarse-graining effect intimately connected to the properties of entanglement entropy. Indeed, this problem can be further related to black holes and thus made more amenable to the holographic gauge/gravity duality as follows.

We have a physical system S (which for concreteness we take to be Schrödinger's cat C), a detector D and the environment E, which both act as encapsulated observers whom we call Wigner's friend (WF) and Wigner (W), respectively. We add a third encapsulated super-observer whom we call Einstein, who observes the joint system $S + D + E$. This is a variant of the Wigner's friend experiment discussed by Susskind [9, 10].

The cat C is as usual in the superposition state of being dead and alive. This is just one qubit that Wigner's friend WF (detector D as an observer) is usually taken to measure. However, the cat C is actually a collection of large N of qubits and WF measures all these qubits.

In the Copenhagen interpretation the measurement of Wigner's friend collapses the state of the cat to one of its 2^N possible states in an irreducible way.

In the many-worlds formulation the measurement of Wigner's friend causes the qubits of the cat C to become entangled with the qubits of the memory states of Wigner's friend. This is the description which Wigner W (the environment E as a Copenhagen observer) should use before he performs his measurement and collapses the state vector of the joint system $C + WF$ (subjective-collapse model). But Wigner W himself, as a many-worlds observer, becomes entangled with the system $C + WF$ when he performs his measurement.

By the stronger ER=EPR conjecture the qubits of the cat C which are entangled with the qubits of Wigner's friend are actually entangled because they are connected by quantum Einstein–Rosen bridges or quantum wormholes.

The weaker form of the ER=EPR uses the fungibility, i.e. the interchangeability, of the entanglement between its various forms [9]. If we assume that the degrees of freedom of a black hole can be represented as a collection of qubits, then by compressing the cat C and Wigner's friend WF into black holes, these black holes will be entangled, and as a consequence there will be quantum Einstein–Rosen bridges between the cat C (the first black hole) and Wigner's friend WF (the second black hole). We will also assume that messages between the cat C and Wigner's friend can be sent to meet in the wormhole.

We can also compress the cat into Hawking radiation and Wigner's friend into a black hole which allows us to the view the joint system $C + WF$ as an evaporating black hole entangled with its Hawking radiation.

Next comes the measurement of Wigner W himself as a Copenhagen observer (the measurement of the environment E itself through decoherence). The Einstein–Rosen bridge is cut at the WF end (because he is the one performing the measurement) and thus messages between the cat C and Wigner's friend WF cannot meet in the wormhole after the completed measurement is done. The snipped ER bridge corresponds to a mixed density matrix and the snipping will cause a firewall between the cat C (viewed as Hawking radiation) and Wigner's friend (viewed as an evaporating black hole), i.e. the information loss as characterized by the release of the firewall is exactly seen as the collapse of the state vector as characterized by the snipped ER bridge. The snipped ER bridge can be made smooth again (corresponding to a unitary process) if Wigner cooperates with his friend, as we will see shortly.

Indeed, with respect to Einstein (the third observer) the joint system Wigner + Wigner's friend + cat is described by a single quantum state. When Wigner performs his measurement on the system Wigner's friend + cat he becomes entangled with it. The system Wigner + Wigner's friend + cat is described by an entangled tripartite state in which the cat is entangled with the union of Wigner and his friend. This is a GHZ state [11].

Thus, when the three systems (Wigner, Wigner's friend and the cat) are compressed into three black holes we obtain an Einstein–Rosen bridge connecting the three which Susskind also calls the GHZ brane [9]. This bridge allows messages between the cat and Wigner's friend to be sent to meet inside the GHZ brane in contradiction with the conclusion of Wigner, using the Copenhagen interpretation, which states that messages cannot be communicated.

The crucial property of the GHZ brane or Einstein–Rosen bridge connecting the three black holes (compressed cat, compressed friend and compressed Wigner) is the fact that it corresponds to a maximally entangled GHZ tripartite state in the same way that the earlier Einstein–Rosen bridge connecting the two black holes (compressed cat and compressed friend) corresponds to a maximally entangled bipartite Bell state. In some sense then we might have called the earlier Einstein–Rosen bridge a Bell brane. The GHZ state is a state in which the union of two of its sub-systems is maximally entangled with the third one (as opposed for example to the W state in which the entanglement is less than maximal).

This property will also lie at the heart of the resolution of the apparent inconsistency between the Copenhagen and the many-worlds interpretations with regard to whether or not there can be messages sent out to meet in the wormhole. As it will turn out, these two interpretations are actually complementary or dual rather than contradictory.

The GHZ brane is thus characterized essentially by maximal GHZ entanglement which is invariant under local unitary transformations. This is a geometric structure localized behind the three horizons of the three entangled black holes which generalizes the Einstein–Rosen bridge between two entangled black holes. The GHZ brane is different from a classical tripartite wormhole which has no GHZ entanglement and thus the two structures can be differentiated. The properties of the GHZ brane encodes the duality between the Copenhagen and many-worlds interpretations.

To construct the GHZ state we start by simply replacing the cat and Wigner's friend by single qubits. The initial state of Wigner is denoted by $|0\rangle$ and his measurement of the state of his friend and the cat will cause him to become entangled with the state of his friend. If he finds his friend in the state $|0\rangle$ (the friend saw a dead cat) he will remain in the state $|0\rangle$, whereas if he finds his friend in the state $|1\rangle$ (the friend saw an alive cat) Wigner will transition to the state $|1\rangle$. The state of Wigner + Wigner's friend + cat is then given by

$$\frac{1}{\sqrt{2}}(|00\rangle|0\rangle + |11\rangle|1\rangle). \tag{4.3}$$

This is the GHZ state for single qubits. It is generally true that when a measurement is performed on one member of an entangled bipartite system a GHZ state is induced. The main two properties of the GHZ state are given by:
- By tracing over any two qubits the density matrix of the third qubit is maximally mixed. This means that the union of any two qubits is maximally entangled with the third qubit.
- By tracing over any one qubit we obtain a separable density matrix of the other two. In other words, the density matrix is a sum of projection operators on unentangled pure states and as a consequence there is no entanglement between any two parties.

This can be generalized to any three complex systems each of which is formed out of collections of a large number of qubits. We obtain then a product of GHZ triplets corresponding to some tensor network.

It is now almost obvious how the discrepancy found between the Copenhagen and the many-worlds interpretations can be resolved. In the Copenhagen interpretation when Wigner measures the state of the system friend + cat the Einstein–Rosen bridge between Wigner's friend and the cat is snipped, because collapse has occurred and entanglement is lost, and as a consequence any messages between the two (Wigner's friend and the cat) cannot be sent to meet inside the wormhole. See figure 4.1.

The many-worlds formulation is appropriate for Einstein when describing the state of Wigner, the friend and the cat. This is given by a GHZ state and indeed because of the separability condition no messages can be sent between any two parties since there is no entanglement between any two parties. So the two interpretations are consistent. However, from this perspective the many-worlds interpretation seems more general than the Copenhagen interpretation because it captures another very important effect due to the preservation of reversibility and entanglement. For Einstein there can still be messages sent between Wigner's friend and the cat if the friend cooperates with Wigner. This is indeed true because of the first property above of the GHZ state which asserts that the union between any two parties is entangled with the third party and as a consequence messages can be sent between the union of any two parties and the third one. Hence the two perspectives are consistent and it is in this sense that the many-worlds interpretation is dual to the Copenhagen interpretation.

The duality between the Copenhagen and many-worlds interpretations guarantees that our description is strictly unitary and as a consequence there can be no measurement problem more than as a coarse-graining effect.

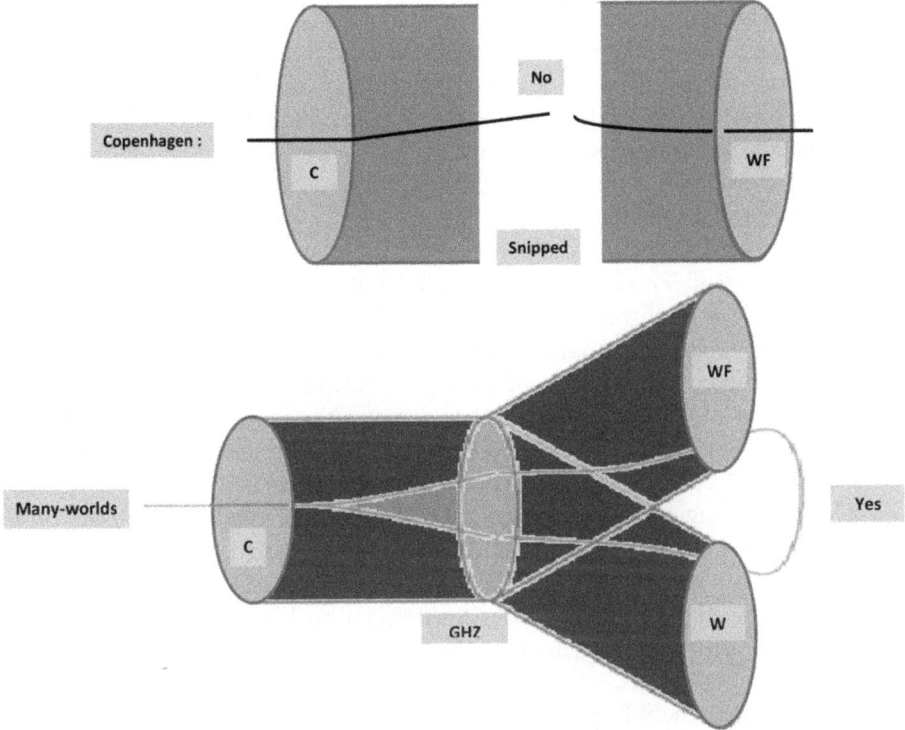

Figure 4.1. The duality between the Copenhagen and many-worlds interpretations.

Indeed, the measurement problem is due mainly to the quantum dualism inherent in the Copenhagen local view of the single-world reality, which fortunately can be mapped to a strictly physicalist view of the many-worlds reality in the same way that the information loss problem is only due to the existence of an event horizon for the local Schwarzschild observer at infinity. The environment plays for the Copenhagen observer the same role as the interior of the black hole for the Schwarzschild observer.

Equally, and perhaps more importantly, there is the possibility of the fungibility of entanglement which allows us the conversion (through unitary transformations) of physical systems, observers and environments into black holes and/or Hawking radiation, which in principle can be made dual to a unitary conformal field theory via the holographic gauge/gravity correspondence. The measurement problem can then be solved through the same means as the information loss problem, at least in principle.

4.3 Quantum dualism

Let us start by remarking that the fundamental problem of the interpretation of quantum mechanics is essentially equivalent to the measurement problem. This can be summarized in the following main points:
- *Copenhagen interpretation*: Unitarity evolution + collapse of the state vector. Measurement: the collapse takes a pure state to a mixed state through decoherence or consciousness or gravity.

- *Many-worlds formalism*: Unitarity evolution + branching. Measurement: branching into coherent worlds (parallel worlds through decoherence).
- *Collapse of the wave function*: Non-unitary collapse (irreversible) is an approximation of unitary branching (reversible). This works because the branches are completely decohered.
- *The Copenhagen/many-worlds correspondence*: A stronger view is that the Copenhagen interpretation provides the local view of reality whereas the many-worlds interpretation provides the manifold view (more on this below).
- *Bell's theorem*: The quantum entangled state does not exist before measurement. This is confirmed by Wheeler's delayed choice.
- *EPR experiment*: Entanglement is spooky action at a distance.
- *Bell's theorem + EPR*: Causal realism or locality must go.
- *Decoherence*: The state of the system + environment is a pure state. However, due to entanglement the state of the system is mixed, given by tracing over the degrees of freedom of the environment which gives a reduced density matrix. Hence, entanglement seems to give rise to collapse. The density matrix undergoes therefore a decrease of information called decoherence through interaction with the environment which is seen as collapse.
- *The entanglement entropy*: The measurement problem is intimately related to the properties of entanglement entropy.

Next, we recall that the information loss problem is another fundamental quantum problem intimately connected to the role of the observer and the properties of entanglement entropy. The main points to remember here are:
- *Black hole information loss*: A correlated entangled pure state near the horizon gives rise to a thermal mixed state outside the horizon.
- *Laws of nature*: Black hole dynamics contradicts one or more of the following laws of nature: unitarity (information conservation), equivalence principle (smoothness of the spacetime manifold), quantum xerox principle (linearity of quantum mechanics), Bekenstein–Hawking entropy (number of states equal the exponential of the surface area of the horizon divided by $4G$), and effective field theory (quantum field theory in weak gravitational backgrounds is valid).
- *Page curve*: The computation of the Page curve starting from first principles will provide, in some precise sense, the mathematical solution of the black hole information loss problem (information starts to get out at the Page time when the entanglement entropy between the interior and the exterior becomes maximal).
- *Black hole complementarity principle*: The information is both reflected at the event horizon (with respect to the external observer) and passed through the horizon toward the singularity (with respect to the infalling observer).
- *Firewall*: A Hawking particle B must be entangled with the early part R_B of the Hawking radiation and with the particle A which went behind the horizon (complementarity). This is, however, forbidden by monogamy. The AMPS resolution is via a loss of entanglement by breaking the entanglement between

B and A. Since A is a high energy mode the breaking will release a firewall and the interior of a black hole is thus not smooth.
- *ER=EPR*: The entanglement between A (behind the horizon) and B (outside the horizon) is possible via an ER bridge connecting them. Thus it resolves the information loss problem by stating that the Hawking radiation is actually connected by ER bridges to itself and also to the black hole horizon and interior.

The hypothesis of 'quantum dualism' is the correspondence observed between the first-person observers of the Copenhagen interpretation and the third-person observers of the many-worlds formalism. This is summarized as follows:
- The collapse is dual to branching and the Copenhagen interpretation is dual to the many-worlds formalism. The many-worlds formalism is an external view (like the curved manifold structure in general relativity) in which the mathematics has an objective reality whereas the Copenhagen interpretation is an internal view (the perceived flatness around every point in a manifold) in which the mathematics has certainly also an objective reality but the representation of reality is only local, i.e. namely a local conscious/zombie observer[1].
- Quantum mechanics is genuinely dualistic (the Copenhagen and von Neumann–Wigner interpretations) and at the same time physicalist (the many-worlds formalism) and the two views are complementary not contradictory, i.e. one can always dualize one into the other consistently.
- This quantum dualism is rooted on the assumptions of:
  - The physical world being quantum mechanical.
  - The von Neumann–Wigner interpretation (quantum dualism, i.e. the independent existence of mind and consciousness).
  - Copenhagen/many-worlds duality (the consciousness of the observer causing collapse can be recast as a unitary description by a super-observer).
- An efficacious consciousness in a single-world as described by Copenhagen quantum mechanics is thus effectively equivalent to a many-worlds scenario where consciousness plays no more than the role it plays in classical mechanics.

The hypothesis of 'quantum dualism' furnishes a unified view of both the measurement problem and the information loss problem allowing us therefore to resolve the former in terms of the latter. The main argument goes as follows:
- The information loss problem in quantum black holes and the measurement problem in quantum mechanics share a unified mathematical structure, namely that an initial pure state is evolved to a final mixed state where the inaccessible environment in quantum mechanics plays exactly the same role as the interior of the event horizon in black holes.

---

[1] In this paper the word 'zombie' refers to the philosophical zombie, which shares with conscious human beings structure and even functional consciousness but lacks subjective qualitative experience (qualia).

- The information loss problem is only a coarse-graining effect in any model of quantum gravity based on the holographic gauge/gravity duality because unitarity is guaranteed in an obvious way.
- Similarly, the complementarity between Copenhagen and many-worlds observers guarantees that our description is strictly unitary and as a consequence there can be no measurement problem more than as a coarse-graining effect. Thus, the measurement problem is due mainly to the quantum dualism inherent in the Copenhagen view of a single-world, which fortunately can be mapped to a strictly physicalist account of the many-worlds view.
- Together with the fungibility of entanglement, which allows us to compress physical systems, observers and environments into black holes and/or Hawking radiations, the measurement problem can be recast as an information loss problem. Thus in principle it can be solved through the same means, i.e. by employing the holographic gauge/gravity duality which maps black holes into conformal field theory.
- Indeed, the gauge/gravity duality provides a framework for a novel interpretation of quantum mechanics in which it is seen that the large N limit of say the M-(atrix) quantum mechanics of BFSS becomes given by classical supergravity around a classical black hole. Thus, in the large N limit the role of the conscious observer decouples from the unitary evolution and there is no measurement problem or collapse of the wave function. The effect of the observer can then be seen as given by a $1/N$ expansion and as such it is also intimately related to the evaporation of the black hole.

Although we have concentrated mainly on the Copenhagen and many-worlds interpretations and on string theory as candidates for quantum gravity, there are so many ingenious interpretations and theories of quantum gravity that it is very difficult for anybody to know them all.

References

[1] Wigner E 2001 Remarks on the mind body question *Philosophical Reflections and Syntheses. The Collected Works of Eugene Paul Wigner. Part B: Historical, Philosophical, and Socio-Political Papers* ed J Mehra vol B/6 (Berlin: Springer)

[2] Wigner E 1963 The problem of measurement *Am. J. Phys.* **31** 6–15

[3] Hameroff S and Penrose R 2014 Consciousness in the Universe: a review of the 'Orch OR' theory *Phys. Life Rev.* **11** 39–78

[4] Maldacena J M 1999 The large N limit of superconformal field theories and supergravity *Int. J. Theor. Phys.* **38** 1113

Maldacena J M 1998 The large N limit of superconformal field theories and supergravity *Adv. Theor. Math. Phys.* **2** 231

[5] Maldacena J M 2003 Eternal black holes in anti-de Sitter *J. High Energy Phys.* **0304** 021

[6] Van Raamsdonk M 2010 Building up spacetime with quantum entanglement *Gen. Rel. Grav.* **42** 2323

van Raamsdonk M 2010 Building up space-time with quantum entanglement *Int. J. Mod. Phys.* D **19** 2429

[7] Hanada M, Hyakutake Y, Ishiki G and Nishimura J 2014 Holographic description of quantum black hole on a computer *Science* **344** 882
[8] Banks T, Fischler W, Shenker S H and Susskind L 1997 M theory as a matrix model: a conjecture *Phys. Rev.* D **55** 5112
[9] Susskind L 2016 Copenhagen vs Everett, teleportation, and ER=EPR *Fortsch. Phys.* **64** 551
[10] Susskind L 2016 ER=EPR, GHZ, and the consistency of quantum measurements *Fortsch. Phys.* **64** 72
[11] Greenberger D M, Horne M A and Zeilinger A 1989 Going beyond Bell's theorem *Bell's Theorem, Quantum Theory, and Conceptions of the Universe* ed M Kafatos (Dordrecht: Kluwer) pp 69–72

IOP Publishing

Philosophy and the Interpretation of Quantum Physics

Badis Ydri

Chapter 5

Black hole interpretation of quantum mechanics

The information loss problem and the measurement problem share the fundamental characterization that an initial pure state is evolved into a final mixed state since there is in both cases a part of the system which is inaccessible (the environment in the case of quantum mechanics and behind the horizon for the case of black holes). However, the information loss problem admits in principle a solution via the holographic gauge/gravity duality. The goal is to exploit this formal analogy (and the analogy as we will argue is even physical) in order to extend the proposed solution for the information loss problem to the measurement problem.

In this chapter a unitary solution inspired by the holographic gauge/gravity duality of the measurement problem is attempted.

5.1 On the complementarity between Copenhagen and many-worlds observers

Quantum mechanics in the standard Copenhagen interpretation [1], as formulated originally by von Neumann [2], is characterized by two processes: process I is the collapse of the wave function which occurs during the quantum measurement and process II is the unitary evolution in time given by the Schrödinger equation.

These two fundamental processes can be succinctly illustrated using the Schrödinger's cat experiment [3] or its extension the Wigner's friend experiment [4]. Thus, Schrödinger is taken to play the role of Wigner's friend who performs a quantum measurement on the state of the joint system cat + atom, whereas Wigner will perform a quantum measurement on the state of the total joint system Schrödinger + cat + atom.

On the one hand, Schrödinger will provide the Copenhagen description of the system cat + atom, which is found to be in a linear coherent superposition of dead-cat/decayed-atom and alive-cat/undecayed-atom until and unless a measurement is performed at which point the state will collapse to one of the said classical alternatives. In other words, Schrödinger provides the local subjective one-person description of reality with direct conscious access to classical pointer states.

On the other hand, the famous/infamous many-worlds description [5] is given by Wigner's account who only sees the linear coherent superposition of sad-observer/dead-cat/decayed-atom and happy-observer/alive-cat/undecayed-atom, i.e. Wigner provides the global objective third-person description of reality who can see the branching worlds directly when a measurement is performed.

These two descriptions are actually complementarity and not mutually exclusive and this extra principle of complementarity should be thought of as extending the usual complementarity principle between Copenhagen first-person observers due to Bohr [1] to include the third-person many-worlds observers.

Of course Nature is seen to be populated only with the first-person local observers of the Copenhagen interpretation but obviously the third-person global observers are not logically prohibited although they are not seen in Nature. Indeed, the fact that such observers are themselves not directly observed in Nature does not rule them out either metaphysically or logically, or even physically.

5.2 On the black hole interpretation of quantum mechanics

The black hole interpretation of quantum mechanics put forward in a sketchy form in the previous chapters relies on strict physicalism [6], i.e. there is no observer-participancy [7] and/or consciousness which is separate from matter [8]. This interpretation employs in a crucial way properties of entanglement entropy and falls within the broad scheme of the Copenhagen interpretation (although it does not treat the collapse as a fundamental process but only as an effective description). It is motivated by the ideas of Susskind in [9] who argues, among other things, for the complementarity relation between the Copenhagen interpretation and the many-worlds formalism using the properties of entanglement entropy and black holes.

Furthermore, the complementarity between the local first-person observer of the Copenhagen interpretation (who sees the collapse of the wave function) and the global third-person observer of the many-worlds formalism (who sees a directly coherent linear superposition) is the analogue of the complementarity between the asymptotic and infalling observers in black holes.

The inaccessible degrees of freedom, with respect to the local asymptotic observer, in the case of the black hole lie behind the horizon, whereas in the case of quantum mechanics the inaccessible degrees of freedom lie behind the Heisenberg cut [10] and thus consist in the degrees of freedom associated with the environment (according to decoherence [11]) or with the consciousness of the local first-person observer (within the Wigner–von Neumann interpretation [4]). Thus, the collapse of the wave function in the quantum measurement problem is the analogue of the information loss problem in black holes since they both result from tracing out the inaccessible degrees of freedom yielding a mixed state.

Thus, we have for the quantum measurement problem (where S is the system, D is the detector and E is the environment)

$$\rho_e = |\Phi_e\rangle\langle\Phi_e| \longrightarrow \rho_r = \rho_{S+D} = \text{Tr}_E|\Psi\rangle\langle\Psi|, \quad |\Psi\rangle = U|\Phi_e\rangle|E_0\rangle. \quad (5.1)$$

whereas for the information loss problem we have (where $\mathcal{S}$ is the singularity)

$$\rho_{\text{in}} = |\psi_{\text{in}}\rangle\langle\psi_{\text{in}}| \longrightarrow \rho_{\text{out}} = \rho_{\text{Hawking}} = \text{Tr}_S|\psi_{\text{out}}\rangle\langle\psi_{\text{out}}|, \quad |\psi_{\text{out}}\rangle = S|\psi_{\text{in}}\rangle. \quad (5.2)$$

However, the information loss problem is only a coarse-graining effect in models of quantum gravity based on the gauge/gravity duality, in particular the holographic AdS/CFT correspondence [12], since unitarity is guaranteed in these cases in an obvious way.

Similarly, the complementarity between the Copenhagen and many-worlds interpretations guarantees that the measurement problem can also be understood as a coarse-graining effect. Together with the fungibility, i.e. the interchangeability of entanglement, which allows us to compress physical systems, observers and environments into black holes and/or Hawking radiations, the measurement problem can be recast as an information loss problem. See for example [9].

The ER=EPR conjecture [13] in which gravitational Einstein–Rosen bridges or wormholes [14] are viewed as Einstein–Podolsky–Rosen–Bell quantum entanglement [15, 16] and vice versa will play an important role. For example, by compressing the cat and Wigner's friend, i.e. Schrödinger into black holes, it is seen by employing the ER=EPR conjecture that these black holes will be entangled, and as a consequence there will be quantum Einstein–Rosen bridges between the cat (first black hole) and Schrödinger (second black hole). We can also compress the cat into Hawking radiation and Schrödinger into a black hole or vice versa which allows us to the view the joint system cat + Schrödinger as an evaporating black hole entangled with its Hawking radiation.

After completed measurement, the Einstein–Rosen bridge is cut at the Schrödinger end (because he is the one performing the measurement) and thus messages between the cat and Schrödinger cannot meet in the wormhole. The snipped ER bridge corresponds to a mixed density matrix and the snipping will cause a firewall between the cat (viewed as Hawking radiation) and Schrödinger (viewed as an evaporating black hole), i.e. the information loss as characterized by the release of the firewall [17] is seen exactly as the collapse of the state vector as characterized by the snipped ER bridge.

In the many-worlds formalism when Wigner performs his measurement on the system Schrödinger + cat he becomes entangled with it. The system Wigner + Schrödinger + cat is described by an entangled tripartite GHZ state (which is a generalization of Bell states to three qubits [16]) in which the cat is entangled with the union of Wigner and his friend [18]. Thus, when the three systems (Wigner, Schrödinger and the cat) are compressed into three black holes we obtain an Einstein–Rosen bridge connecting the three which Susskind also calls the GHZ brane [9]. This bridge allows messages between the cat and Wigner's friend to be sent to meet inside the GHZ brane in contradiction with the conclusion, using the Copenhagen interpretation, which states that messages cannot be communicated.

However, because of the separability property of entanglement entropy no messages in the GHZ state can be sent between any two parties since there can be no entanglement between any two of them. So the two interpretations are consistent.

However, from this perspective the many-worlds formalism seems more general than the Copenhagen interpretation because it captures another very important effect due to the preservation of reversibility and entanglement. There can still be messages sent between Schrödinger and the cat if Schrödinger cooperates with Wigner since the union between any two parties in the GHZ state is entangled with the third party, and as a consequence messages can be sent between the union of any two of them and the third one. Hence the two perspectives are consistent and it is in this sense that the many-worlds formalism is complementary to the Copenhagen interpretation. Thus, in principle the measurement problem can be solved (or understood) through the same means used to understand the black hole information loss problem, i.e. by employing the gauge/gravity duality which maps black holes into gauge theory.

0A confirmation, using non-perturbative Monte Carlo methods, of the fact that black holes follow the predictions of the gauge/gravity correspondence is carried out in [19], where the gauge theory is given in their case by the celebrated M-(atrix) quantum mechanics, also known as the BFSS matrix model [20], whereas on the gravity side we have string theory around the black 0-brane (which is a black hole in ten dimensions).

Thus, the gauge/gravity duality provides a framework for a novel interpretation of quantum mechanics in which it is seen that the large N limit of M-(atrix) quantum mechanics becomes given by classical supergravity around a classical black hole given by the 0-brane configuration. The wave function of the system or more precisely its path integral is given exactly by the classical supergravity action, namely

$$\Psi = Z_{\text{BFSS}} = \exp(-S_{\text{SUGRA}}). \tag{5.3}$$

In other words, there can be no linear coherent superposition of classical states in this limit since the wave function is already fully decohered. Thus, in the large N limit the role of the observer decouples from the unitary evolution and there is no measurement problem or collapse of the wave function.

The supergravity solution is in fact a saddle point of the superstring path integral and in general the BFSS matrix model defines the dual gauge theory of the full superstring theory around the black 0-brane. The effect of the observer in the BFSS quantum mechanics side can then be seen as being given by a $1/N$ expansion and as such it is intimately related to the evaporation of the black hole, which we will now explain.

First, let us mention for completeness that in the gauge/gravity correspondence [12] the two parameters N (rank of the gauge group) and g_{YM} (the gauge coupling constant) on the gauge theory (quantum mechanics) side are related to the two parameters l_s (string length) and g_s (string coupling constant) on the string theory (black hole and gravity) side as follows:

1. The gauge theory in the limit $N \longrightarrow \infty$ and $\lambda \longrightarrow \infty$ should be equivalent to classical type IIA supergravity around the 0-brane spacetime. Recall that $\lambda = g_{\text{YM}}^2 N$ is the 't Hooft coupling constant [21] and $\alpha' = l_s^2$ is the inverse of the string tension.
2. The gauge theory with $1/N^2$ corrections should correspond to quantum loop corrections, i.e. corrections in g_s, on the gravity/string side.
3. The gauge theory with $1/\lambda$ corrections should correspond to stringy corrections, i.e. corrections in l_s, corresponding to the fact that the degrees of freedom on the gravity/string side are really strings and not point particles.

5.3 On observers in quantum mechanics and the evaporation of black holes

In the remainder of this chapter we will attempt to show that the effect of the observer in the quantum mechanics of M-(atrix) theory is intimately related to the evaporation of the black hole of the dual gravity theory of the black 0-brane.

First, the black hole in this theory is the black 0-brane solution which is a bound state of N D0-branes (or D-particles) connected by open strings. Indeed, on the gauge theory side the degrees of freedom are given by nine $N \times N$ Hermitian matrices X_I whose diagonal elements describe the positions of the N D-particles forming the black hole, whereas the off-diagonal elements are fields describing the open strings stretching between these D-particles.

Next, the black 0-brane is a solution of type IIA superstring theory which, when lifted to 11 dimensions, becomes the M-wave solution of M-theory which is a purely geometrical object [22]. Thus, the relevant effective action of M-theory can be computed by concentrating only on the graviton field and demanding local supersymmetry. The effective action of type IIA superstring theory is then obtained via dimensional reduction. Analogously, the near-horizon geometry of the black 0-brane with quantum gravity corrections included is obtained by dimensionally reducing the near-horizon geometry of the M-wave solution.

The thermodynamical properties (temperature, entropy, energy and specific heat) of the black 0-brane can then be computed in the standard way. For example, the (Hawking) temperature and the specific heat are found to be given by (where $\tilde{U}_0$ indicates the classical horizon in units of 't Hooft coupling $\lambda^{1/3}$)

$$\tilde{T} = a_1 \tilde{U}_0^{5/2}(1 + \epsilon a_2 \tilde{U}_0^{-6}). \tag{5.4}$$

$$\frac{1}{N^2}\frac{d\tilde{E}}{dT} = \frac{9a_3}{5}\tilde{T}^{9/5} - \frac{3\epsilon a_3 a_4}{5}\tilde{T}^{-3/5}. \tag{5.5}$$

The numerical coefficients a_i can be found in [22] and $\epsilon \sim 1/N^2$. Thus, the specific heat can become negative at low temperatures which means that the black 0-brane behaves as an evaporating Schwarzschild black hole. This instability is obviously removed in the limit $N \longrightarrow \infty$.

This instability of the black hole solution (which corresponds physically to Hawking radiation) is associated with the divergence of the eigenvalues of the matrices X_I as N becomes small. In the BFSS matrix model, this is related to the problem of flat directions or commuting matrices which have zero action in the path integral. A powerful order parameter which exhibits this behavior is given by the so-called extent of space defined by

$$R^2 = \frac{1}{N\beta}\int_0^\beta dt \sum_{I=1}^{9} X_I(t)^2. \tag{5.6}$$

The period β of the imaginary time t (the time parameter which appears in the BFSS action) is related to the Hawking temperature (5.4) in the usual way, i.e. $\beta = 1/\tilde{T}$.

It is observed in Monte Carlo simulations that the distribution of R^2 presents a peak (bound state) and a run-away tail (Hawking instability) [19]. The black hole for small N is therefore only a metastable bound state of the D0-branes and quantum gravity is acting as a destabilizing effect.

This instability can be approximated as follows. By making in the BFSS matrix model the two approximations of (i) a quenched determinant (bosonic model) and (ii) a large number of spatial dimensions d (the Yang–Mills term is approximated by a mass term) we observe at low temperatures (for which the holonomy is exponentially suppressed) that the BFSS matrix model is a scalar field theory given by the action (see [23] and references therein)

$$S = N \int_0^\infty dt \, \text{Tr}\left[\frac{1}{2}(\partial_t X_I)^2 + \frac{m^2}{2} X_I^2\right], \quad m = d^{1/3}\lambda^{1/3}. \tag{5.7}$$

The eigenvalue distribution of any one of the X_I is then given by a Wigner semicircle law with a radius given by

$$R_\lambda = \sqrt{\frac{2}{m}} = \sqrt{\frac{2}{d^{1/3}\lambda^{1/3}}}. \tag{5.8}$$

This radius can also be expressed in terms of the expectation value of the extent of space R^2 by the relation

$$R_\lambda = \frac{4}{d}\langle R^2 \rangle. \tag{5.9}$$

In summary, as long as we are permitted to use the bosonic approximation (which seems reasonable at low temperatures) and the large dimension expansion (which is quite justified since $d = 9$), it follows that the holonomy of the gauge field is exponentially suppressed at low temperatures and the dynamics of the BFSS quantum mechanics is given effectively by (5.7), which describes a matrix harmonic oscillator. As a consequence, we have as follows:

1. Strictly speaking the results (5.8) and (5.9) are an $N = \infty$ effect but it is also valid for sufficiently large values of N. This clearly vanishes in the limit $\lambda \longrightarrow \infty$ corresponding to the classical black 0-brane solution associated with the saddle point (5.3). This represents the fact that in the classical regime the D-particles are absolutely coincident and the black hole does not evaporate Hawking radiation.

2. The eigenvalue distribution of each of the matrix coordinates X_I, for large but finite values of N and λ, is given by a Wigner semicircle law with radius (5.8) which increases as we decrease λ. This represents the fact that the D-particles are spread over large distances in the quantum regime thus modeling the Hawking radiation.

   The resulting extent of space R^2 given by equation (5.9) should then be thought of as an approximation of the distribution of the extent of space in the full model which presents a peak (bound state) and a run-away tail (Hawking instability) as we decrease N [19].

3. The wave function of the system at low temperatures is the ground state of the matrix harmonic oscillator (or a superposition which is largely dominated by the ground state). It is expected that this picture should hold in the full BFSS model, i.e. the wave function is expected to be largely dominated by the ground state of the system at low temperatures. At high temperature the partition function is given by the classical result which is identified with the saddle point (5.3).

Hence there is no linear coherent superposition in these limits (or they are overwhelmingly dominated by a preferred state) and thus, as before, the role of the observer decouples from the unitary evolution and there is no measurement problem or collapse of the wave function. The radiated modes away from the black hole are seen to lie at the root of the superposition principle and thus they lie at the root of the measurement problem.

References

[1] Bohr N 1963 *Essays 1958–62 on Atomic Physics and Human Knowledge* (New York: Wiley)
Bohr N 1949 Discussions with Einstein on epistemological problems in atomic physics *Albert Einstein: Philosopher-Scientist* (Cambridge: Cambridge University Press), see also [7, 10]

[2] von Neumann J 1955 *Mathematical Foundations of Quantum Mechanics* 1st edn (Princeton, NJ: Princeton University Press)

[3] Trimmer J D 1980 The present situation in quantum mechanics: a translation of Schrödinger's 'cat paradox' paper *Proc. Am. Phil. Soc.* **124** 323–38

[4] Wigner E 2001 Remarks on the mind body question *Philosophical Reflections and Syntheses. The Collected Works of Eugene Paul Wigner. Part B: Historical, Philosophical, and Socio-Political Papers* ed J Mehra vol B/6 (Berlin: Springer)

[5] Everett H 1957 Relative state formulation of quantum mechanics *Rev. Mod. Phys.* **29** 454–62, see also [7]

[6] Kim J 2005 *Physicalism or Something Near Enough* (Princeton, N J: Princeton University Press)

[7] Wheeler J A 1990 Information, physics, quantum: the search for links *Complexity, Entropy, and the Physics of Information* ed W H Zurek (Redwood City, CA: Addison-Wesley)

[8] Stapp H 2007 Quantum mechanical theories of consciousness *The Blackwell Companion to Consciousness* ed M Velmans and S Schneider (Berlin: Springer)
Stapp H P 2009 *Mind, Matter and Quantum Mechanics The Frontiers Collection* (Berlin: Springer)

[9] Susskind L 2016 Copenhagen vs Everett, teleportation, and ER=EPR *Fortsch. Phys.* **64** 551
Susskind L 2016 ER=EPR, GHZ, and the consistency of quantum measurements *Fortsch. Phys.* **64** 72

[10] Heisenberg W 2011 Is a deterministic completion of quantum mechanics possible? transl. by Crull E and Bacciagaluppi G https://halshs.archives-ouvertes.fr/halshs-00996315

[11] Zeh H D 1970 On the interpretation of measurement in quantum theory *Found. Phys.* **1** 69–76

[12] Maldacena J M 1999 The large N limit of superconformal field theories and supergravity *Int. J. Theor. Phys.* **38** 1113
Maldacena J M 1998 The large N limit of superconformal field theories and supergravity *Adv. Theor. Math. Phys.* **2** 231

[13] Maldacena J and Susskind L 2013 Cool horizons for entangled black holes *Fortsch. Phys.* **61** 781
[14] Einstein A and Rosen N 1935 The particle problem in the general theory of relativity *Phys. Rev.* **48** 73
[15] Einstein A, Podolsky B and Rosen N 1935 Can quantum mechanical description of physical reality be considered complete? *Phys. Rev.* **47** 777–80
[16] Bell J S 1964 On the Einstein–Podolsky–Rosen paradox *Physics* **1** 195
[17] Almheiri A, Marolf D, Polchinski J and Sully J 2013 Black holes: complementarity or firewalls? *J. High Energy Phys.* **1302** 062
[18] Greenberger D M, Horne M A and Zeilinger A 1989 A going beyond Bell's theorem *Bell's Theorem, Quantum Theory, and Conceptions of the Universe* ed M Kafatos (Dordrecht: Kluwer) pp 69–72
[19] Hanada M, Hyakutake Y, Ishiki G and Nishimura J 2014 Holographic description of quantum black hole on a computer *Science* **344** 882
[20] Banks T, Fischler W, Shenker S H and Susskind L 1997 M theory as a matrix model: a conjecture *Phys. Rev.* D **55** 5112
[21] 't Hooft G 1974 A planar diagram theory for strong interactions *Nucl. Phys.* B **72** 461
[22] Hyakutake Y 2014 Quantum near-horizon geometry of a black 0-brane *Prog. Theor. Exp. Phys.* **2014** 033B04
[23] Ydri B 2020 Two approaches to quantum gravity and M-(atrix) theory at large number of dimensions arXiv: 2007.04488

IOP Publishing

Philosophy and the Interpretation of Quantum Physics

Badis Ydri

Chapter 6

On quantum logic and quantum metaphysics

As we will see in this chapter 'quantum logic' is a theory construed as an interpretation of quantum mechanics, although for some it is a theory which attempts to go beyond. In fact quantum logic may even be understood as nothing more than an interpretation of the Copenhagen interpretation. A much stronger perspective along this line will be taken up in the next chapter.

Boolean algebras, which are relevant to classical logic, are a special case of orthocomplemented modular lattices, which are relevant to quantum logic, where the distributive law in classical logic is replaced with the modular law in quantum logic. The goal in this chapter is to define all these terms carefully.

In this chapter we will also discuss in detail the Kochen–Specker theorem which can be viewed as a manifestation of the more fundamental Bell's theorem.

6.1 Boolean algebras

Boolean algebras were introduced by George Boole as the first application of algebraic methods to logic and hence Boolean algebras can be thought of as algebras of logic.

6.1.1 Definition

We define a structure

$$(\mathcal{S}, \wedge, \vee, \neg, \mathbf{0}, \mathbf{1}) \tag{6.1}$$

by a set $\mathcal{S}$ together with two binary operations **AND** and **OR** denoted by $\wedge$ and $\vee$, one unary operation **NOT** denoted by $\neg$ and two distinct elements **1** (the unit element) and **0** (the zero element) satisfying the following five properties:

- *Associativity*: For any x, y and z in $\mathcal{S}$ we have

$$x \wedge (y \wedge z) = (x \wedge y) \wedge z, \quad x \vee (y \vee z) = (x \vee y) \vee z. \tag{6.2}$$

- *Commutativity*: For any x, y in S we have
$$x \wedge y = y \wedge x, \quad x \vee y = y \vee x. \tag{6.3}$$

- *Absorption*: For any x, y in S we have
$$x \wedge (x \vee y) = x, \quad x \vee (x \wedge y) = x. \tag{6.4}$$

- *Identity*: For any x in S we have
$$x \wedge \mathbf{1} = x, \quad x \vee \mathbf{0} = x. \tag{6.5}$$

- *Distributivity*: For any x, y and z in S we have
$$x \wedge (y \vee z) = (x \wedge y) \vee (x \wedge z), \quad x \vee (y \wedge z) = (x \vee y) \wedge (x \vee z). \tag{6.6}$$

- *Complementation*: For any x in S we have
$$x \wedge (\neg x) = \mathbf{1}, \quad x \vee (\neg x) = \mathbf{0}. \tag{6.7}$$

A partial ordering on S exists and is defined by the relation
$$x \leqslant y \text{ iff } x \wedge y = x. \tag{6.8}$$

This is a partial order with a least element or bottom given by $\mathbf{0}$ and a greatest element or top given by $\mathbf{1}$. The meet $x \wedge y$ (or greatest lower bound) and the join $x \wedge y$ (or least upper bound) are the infimum and supremum with respect to the ordering relation $\leqslant$, respectively. Thus $x \wedge y$ is the greatest element of S lying below both x and y, while $x \vee y$ is the least element of S lying above both x and y.

Then $(S, \wedge, \vee, \neg, \mathbf{0}, \mathbf{1})$ (or S for short) is a Boolean algebra [1]. In Boolean algebras we also have the following two laws:

- *Law of double complementation*: For any x in S we have
$$\neg(\neg x) = x. \tag{6.9}$$

- *De Morgan's law*: For any x, y in S we have
$$\neg(x \wedge y) = \neg x \vee \neg y, \quad \neg(x \vee y) = \neg x \wedge \neg y. \tag{6.10}$$

Furthermore from these two laws it follows that in any Boolean algebra the operation **AND/OR** is definable in terms of the operations **OR/AND** and **NOT** by employing the equations
$$x \wedge y = \neg(\neg x \vee \neg y), \quad x \vee y = \neg(\neg x \wedge \neg y). \tag{6.11}$$

6.1.2 Boolean algebras as lattices

Lattices
A lattice $\mathcal{L}$ is a set $\mathcal{S}$ with a partial ordering relation $\leqslant$ such that for any x and y in $\mathcal{S}$ there exists a join (or supremum) $x \vee y$ and a meet (or infimum) $x \wedge y$. The ordering is defined by

$$x \leqslant y \iff x \wedge y = x \iff x \vee y = y. \tag{6.12}$$

The properties of associativity, commutativity and absorption given by equations (6.2), (6.3) and (6.4) can be shown to hold in the lattice $\mathcal{L} = (\mathcal{S}, \leqslant)$.

Distributive lattices
If furthermore equation (6.6) holds the lattice $\mathcal{L} = (\mathcal{S}, \leqslant)$ is called distributive.

Complemented lattices
A lattice $\mathcal{L} = (\mathcal{S}, \leqslant)$ is complemented if there exists an identity **1** (greatest element) with respect to the meet operation $\wedge$ and an identity **0** (least element) with respect to the join operation $\vee$ given by equation (6.5), and if for each x in $\mathcal{S}$ there exists y in $\mathcal{S}$ such that

$$x \wedge y = \mathbf{1}, \quad x \vee y = \mathbf{0}. \tag{6.13}$$

In other words, $y = \neg x$.

Three final remarks are in order:
- The only complemented lattice which is totally ordered is trivial.
- In any distributive lattice every element has at most one complement.
- A Boolean algebra is a complemented distributive lattice.

6.1.3 Example 1: The minimal algebra

The simplest Boolean algebra is given by the set $\mathbf{2} = \{0, 1\}$ where the binary digits (bits) 0 and 1 denote truth values *false* and *true*, respectively. The basic operations of Boolean algebras in this case are then:
- **AND** is the *conjunction*. Thus the proposition $x \wedge y$ is true only if both propositions x and y are true otherwise it is false.
- **OR** is the *inclusive disjunction*. Thus the proposition $x \vee y$ is false if both propositions x and y are false otherwise it is true.
- **NOT** is the *negation*. The proposition $\neg x$ is true if x is false and vice versa.

The values $x \wedge y$, $x \vee y$ and $\neg x$ can be expressed using truth tables and Venn diagrams. This Boolean algebra is called the minimal algebra and it is very important in logic and computing.

The ordering relation $(x \leqslant y \iff x \wedge y = x)$ on the set $\mathbf{2} = \{0, 1\}$ is given explicitly by the relations

$$0 \leqslant 0, \quad 0 \leqslant 1, \quad 1 \leqslant 1. \tag{6.14}$$

This ordering relation is actually equivalent to the *material conditional* $\longrightarrow$ given by the following truth table:

$$
\begin{aligned}
x = 0, \quad y = 0, \quad & x \longrightarrow y = 1 \\
x = 0, \quad y = 1, \quad & x \longrightarrow y = 1 \\
x = 1, \quad y = 1, \quad & x \longrightarrow y = 1.
\end{aligned}
\tag{6.15}
$$

The material conditional $x \longrightarrow y$ is false only if $x = 1$ and $y = 0$ corresponding to the absent ordering $1 \leqslant 0$, namely

$$
1 \leqslant 0 \iff x = 1, \quad y = 0, \quad x \longrightarrow y = 0.
\tag{6.16}
$$

The material conditional $x \longrightarrow y$ does not correspond to logical consequence or implication which is denoted by $\Rightarrow$, nor does it entail any causal relationship between the antecedent x and the consequent y (which is a bit paradoxical). It simply states that: 'If x is true, then y is also true'. It is equivalent to the proposition $\neg x \wedge y$.

This can be understood better if we use sets or Venn diagrams. The **AND, OR, NOT** and material conditional correspond on sets to the set-theoretical intersection $\cap$, union $\cup$, complement $-$ and union $\subseteq$, respectively. Thus the ordering relation $(x \leqslant y \iff x \wedge y = x)$ is actually $(x \subseteq y \iff x \cap y = x)$, i.e. x is a subset of y. Hence if a point belongs to x it will necessarily also belong to y (this is the meaning of the material conditional). The equivalent form $\neg x \vee y$ of the material conditional becomes using sets $-x \cup y$. Clearly then the statement 'a point belongs to x but it does not belong to y' is false (this is equation (6.16)). All other possibilities are true: 'a point belongs to x and belongs to y', 'a point does not belong to x and belongs to y' and 'a point does not belong to x and does not belong to y' corresponding to (6.15). The last two cases are difficult to understand but a little contemplation allows us to appreciate that if x is false (the point does not belong to x) then we cannot say anything about the truth value of y and since the material conditional is not contradicted it should be considered true [2].

6.1.4 Example 2: Power sets under inclusion

The most important example of a Boolean algebra appears in set theory when considering the power set $P(X)$ of a set X, i.e. the set of all subsets of X. The power set $P(X)$ under inclusion $\subseteq$ is a distributive lattice, i.e. $\subseteq$ defines a partial ordering on $P(X)$, whereas the union $\cup$ and the intersection $\cap$ play the role of the join $\vee$ and the meet $\wedge$ operations. Indeed, $A \cup B$ is the supremum of A and B since $A \subseteq A \cup B$ and $B \subseteq A \cup B$ and furthermore if $A \subseteq C$ and $B \subseteq C$ then $A \cup B \subseteq C$. Similarly, $A \cap B$ is the infimum of A and B since $A \cap B \subseteq A$ and $A \cap B \subseteq B$ and furthermore if $C \subseteq A$ and $C \subseteq B$ then $C \subseteq A \cap B$.

The identity **1** with respect to the intersection or meet operation $\wedge = \cap$ is the whole set X whereas the identity **0** with respect to the union or join operation $\vee = \cup$ is the empty set $\emptyset$.

The operation of complementation is precisely the set-theoretic complementation with respect to X, i.e. the complement $\neg A$ of a set $A \in P(X)$ is the set of all elements of X which are not in A. $\neg A$ is also a set in $P(X)$.

The power set $P(X)$ under inclusion $\subseteq$ (which acts as the material conditional $\longrightarrow$ here) is a complemented distributive lattice. We have then

$$\begin{aligned} \leqslant &\leftrightarrow \subseteq \\ \wedge &\leftrightarrow \cap \\ \vee &\leftrightarrow \cup \\ \mathbf{1} &\leftrightarrow X \\ \mathbf{0} &\leftrightarrow \varnothing \\ \neg &\leftrightarrow - = \text{complementation with respect to } X. \end{aligned} \qquad (6.17)$$

The complemented distributive lattice $(P(X), \subseteq)$ is actually a Boolean algebra given by the structure

$$(P(X), \cap, \cup, -, X, \varnothing). \qquad (6.18)$$

6.1.5 Example 3: Classical propositional calculus and Lindenbaum–Tarski algebra

In this section we follow [3, 4].

Classical propositional calculus
We consider a propositional language $\mathcal{L}$ consisting of sentences constructed from a basic alphabet with the following six letters:
- $\neg$
- $\longrightarrow$
- p
- $+$
- $($
- $)$

The first two letters $\neg$ and $\longrightarrow$ are called propositional connectives (which are pronounced 'not' and 'implies', respectively), whereas the letter p followed by any finite number of $+$ such as $p, p^+, p^{++}, p^{+++}\ldots$ is a propositional variable. These propositional variables are called sentences and they are denoted by x. We can also form sentences using the propositional connectives such as $\neg x$ (negation of x) and $x \longrightarrow y$ (x implies y).

We can use the two propositional connectives $\neg$ and $\longrightarrow$ to define three more propositional connectives $\wedge$, $\vee$ and $\iff$ given by

$$\begin{aligned} (x \vee y) &= ((\neg x) \longrightarrow y), \quad x \wedge y = (\neg(x \longrightarrow (\neg y))), \\ (x \iff y) &= ((x \longrightarrow y) \wedge (y \longrightarrow y)). \end{aligned} \qquad (6.19)$$

The parentheses $(,)$ can be dropped if no ambiguities may result.

The axioms in the propositional language $\mathcal{L}$ are given by the following four sentences:
- $x \vee x \longrightarrow x$,
- $x \longrightarrow x \vee y$,
- $x \vee y \longrightarrow y \vee x$,
- $(x \longrightarrow y) \longrightarrow (z \vee x \longrightarrow z \vee y)$.

The modus ponens is a rule of inference which states that if x and y are two sentences in $\mathcal{L}$ such that x and $x \longrightarrow y$ are theorems then y is also a theorem.

A sentence x in $\mathcal{L}$ is a theorem if and only if there exists a finite sequence of sentences which terminates with x such that each sentence in the sequence is either (i) an axiom or (ii) is obtained by applying modus ponens to a pair of preceding sentences. This finite sequence of sentences is called the deduction of x. If the sentence x in $\mathcal{L}$ is a theorem we write

$$\vdash x. \tag{6.20}$$

This is read 'x is provable in $\mathcal{L}$'.

Let A be a set of sentences of $\mathcal{L}$. A sentence x in $\mathcal{L}$ is said to be deducible from A if and only if there exists a finite sequence of sentences which terminates with x such that each sentence in the sequence is either (i) an axiom, (ii) is in A or (iii) is obtained by applying modus ponens to a pair of preceding sentences. This finite sequence of sentences is called the deduction of x from A. If the sentence x in $\mathcal{L}$ is deducible from A we write

$$A \vdash x. \tag{6.21}$$

The set A is said to be inconsistent if $A \vdash x \wedge \neg x$ for some sentence x otherwise it is said to be consistent.

A semantic interpretation of the above syntactical structure of the propositional language $\mathcal{L}$ is given by a truth valuation v which is a mapping from the set of all sentences of $\mathcal{L}$ into the set $\{0, 1\}$ which assigns each sentence of $\mathcal{L}$ a truth value 0 (false) or 1 (true). This mapping (truth valuation) satisfies

$$v(\neg x) = 1 \iff v(x) = 0. \tag{6.22}$$

$$v(x \longrightarrow y) = 1 \iff v(x) = 0 \text{ and } v(y) = 1. \tag{6.23}$$

If x is a sentence such that $v(x) = 1$ for all truth valuations v then we call x a tautology and we write

$$\vDash x. \tag{6.24}$$

On the other hand, if x is a sentence such that $v(x) = 0$ for all truth valuations v then we call x an absurdity and we write

$$\vDash \neg x. \tag{6.25}$$

Lindenbaum–Tarski algebra

Let $\mathcal{L}$ be a propositional language and let S be the set of all sentences of $\mathcal{L}$. We define a relation $\equiv$ on the set S by

$$x \equiv y \text{ if and only if } \vdash x \iff y. \tag{6.26}$$

In other words, the sentence x is the same as the sentence y if the sentence 'x is equivalent to y' is provable in $\mathcal{L}$.

The relation $\equiv$ is an equivalence relation. The equivalence class of the sentence x is denoted by $|x|$ while the set of all equivalence classes is denoted by B, i.e.

$$B = \{|x|: x \in S\}. \tag{6.27}$$

We define an ordering relation $\leqslant$ on B by

$$|x| \leqslant |y| \text{ if and only if } \vdash x \longrightarrow y. \tag{6.28}$$

By employing the same symbols, the join $\vee$, the meet $\wedge$ and the complementation $\neg$ on B are defined by

$$|x| \vee |y| = |x \vee y|, \quad |x| \wedge |y| = |x \wedge y|, \quad \neg|x| = |\neg x|. \tag{6.29}$$

The tautology and the absurdity are defined by $|x| = \mathbf{1}$ and $|\neg x| = \mathbf{0}$, where the sentence x is any theorem in $\mathcal{L}$.

Thus $\equiv$ is a congruent relation and B is a quotient algebra. In fact, the equivalence classes $|x|$ of B act as elements of a Boolean algebra called the Lindenbaum–Tarski algebra of $\mathcal{L}$ which is defined by

$$(B, \wedge, \vee, \neg, \mathbf{1}, \mathbf{0}). \tag{6.30}$$

This is arguably the most important Boolean algebra.

6.1.6 Example 4: The logical structure of the classical theory

Newtonian mechanics is a classical theory from both the physical point of view as well as from the logical point of view. It is based on the Boolean algebra of the power set $P(\Sigma)$, where the set Σ is the phase space of generalized coordinates q_i and conjugate momenta p_i. The physical states in Newtonian mechanics are points $x = (q_i, p_i)$ in this phase space Σ whereas physical properties or observables are represented by real functions on Σ which are commuting, i.e. they can be measured simultaneously. Physical events correspond therefore to subsets X of Σ.

More precisely, let S be a classical physical system with a phase space Σ and let $\mathcal{O}_i: \Sigma \longrightarrow \Omega_i$ be some observables on S with observation spaces Ω_i, respectively. The observables $\mathcal{O}_1, \mathcal{O}_2, \ldots, \mathcal{O}_n$ can all be measured simultaneously with an observation space given by the Cartesian product $\Omega_1 \times \Omega_2 \cdots \times \Omega_n$. Every subset ω of $\Omega_1 \times \Omega_2 \cdots \times \Omega_n$ defines a proposition P about the physical system, i.e. some physical quantities of the physical system has certain values. For example, P could be the proposition 'the values of the generalized coordinates q_i lie in the intervals I_i and the values of the conjugate momenta p_i lie in the intervals J_i'.

We have then [5]:

- P: The measurements of the observables $\mathcal{O}_1, \mathcal{O}_2, \ldots, \mathcal{O}_n$ lie in $\omega \in \Omega_1 \times \Omega_2 \cdots \times \Omega_n$ when the state of the physical system S is $x = (q_i, p_i)$.

The propositions P which are called experimental propositions are then correlated with subsets X of Σ given by

$$X = \{x \in \Sigma: (\mathcal{O}_1(x), \mathcal{O}_2(x), \ldots, \mathcal{O}_n(x)) \in \omega\}. \tag{6.31}$$

In other words, $X \in \Sigma$ is the set of states x (pure states of maximal information) of the physical system S such that the experimental proposition P is verified when S is in the state $x = (q_i, p_i)$. When a pure state x is in X we say that the physical system S in state x verifies X and the corresponding experimental proposition P. Hence, physical events are indeed subsets X of the phase space Σ. However, the map between experimental propositions P and physical events (subsets of the phase space) X is in fact many-to-one.

The Boolean algebra of interest in this case is then given by the structure [6]:

$$(F(\Sigma), \subseteq, \cap, \cup, \neg, \mathbf{1}, \mathbf{0}), \tag{6.32}$$

where:
- $F(\Sigma)$ is the set of all measurable subsets of Σ (not the full power set $P(\Sigma)$).
- $\subseteq, \cap, \cup$ are the inclusion, the intersection and the union of set theory.
- The identity $\mathbf{1}$ with respect to $\cap$ is the phase space itself Σ.
- The identity $\mathbf{0}$ with respect to $\cup$ is the empty state $\emptyset$.
- $\neg$ is the set-theoretic complementation with respect to Σ, i.e. $\neg = \overline{}$.

The semantic behavior of this Boolean algebra is given by:
- The intersection $\cap$ is the conjunction $\wedge$: a state x verifies $X \cap Y \iff x \in X \cap Y \iff x \in X$ and $x \in Y$.
- The union $\cup$ is the disjunction $\vee$: a state x verifies $X \cup Y \iff x \in X \cup Y \iff x \in X$ or $x \in Y$.
- A state x verifies a negation $\neg X \iff x \notin X \iff x$ does not verify X.
- The ordering relation: the inclusion $\subseteq$ acts here as the material conditional $\longrightarrow$.

When we go from classical mechanics to statistical mechanics a probability measure is added to the theory which can be described by a classical Kolmogorovian probability.

This Boolean algebra is a distributive lattice where the two main laws of Aristotelian logic holds:
- *The principle of excluded middle*: Either the proposition P or its negation $\neg P$ is true, i.e. pure states semantically decide any event.
- *The principle of non-contradiction*: Both propositions P and $\neg P$ cannot be simultaneously true.

This reflects the fact that in classical physics meaning is unambiguous due to the fact that pure states are states of maximal information and hence yield logical completeness [7].

As we will see, in quantum mechanics the pure states, which are also states of maximal information, do not yield logical completeness and meaning remains ambiguous due to the different structure of the Hilbert space which replaces the structure of the phase space and also due to the principle of linear superposition which allows a state to contain both the property and its negation simultaneously.

6.2 The logical structure of the quantum theory

Quantum logic was initiated by Birkhoff and von Neumann in their seminal paper 'The logic of quantum mechanics' [8] which appeared in 1936 with a primary motivation toward the problem of interpretation of quantum mechanics. In this quantum logic the Boolean algebra of events defined on the classical phase space is replaced by the lattice (non-Boolean, non-distributive and orthocomplemented) of projection operators on the Hilbert space. This quantum logic underlies the Born's rule and quantum probability theory (the quantum states define probability measures on the lattice) in the same way that the Boolean algebra of the classical phase space underlies Kolmogorovian probability theory. In the following we will summarize the main points of their theory.

6.2.1 Hilbert space quantum mechanics

Hilbert space

A Hilbert space H is a linear vector space over the complex numbers $\mathbb{C}$ equipped with an inner or scalar product denoted $(,)$ satisfying:

- The inner product is symmetric under complex conjugation

$$(x, y) = (y, x)^*, \quad \forall x, y \in H. \tag{6.33}$$

- The inner product is linear in its second argument and therefore it is conjugate linear in its first argument, namely

$$(z, \alpha x + \beta y) = \alpha(z, x) + \beta(z, y),$$
$$(\alpha x + \beta y, z) = \alpha^*(x, z) + \beta^*(y, z), \quad \forall x, y, z \in H. \tag{6.34}$$

- The inner product is positive definite, namely

$$(x, x) \geq 0, \quad \forall x \in H. \tag{6.35}$$

$$(x, x) = 0 \text{ if and only if } x = 0. \tag{6.36}$$

The Hilbert space is a complete or Cauchy metric space. Indeed, any inner product space is a metric space with the norm and distance functions defined by

$$\|x\| = \sqrt{(x, x)}, \quad d(x, y) = \|x - y\|. \tag{6.37}$$

This metric space is known as a pre-Hilbert space. The Hilbert space is a metric space which is also complete, i.e. every Cauchy sequence of points in H converges with respect to the distance function to an element in H. This can be characterized as follows. We consider the series consisting of vectors x_n in H given by

$$\sum_{n=0}^{N} x_n, \quad N \longrightarrow \infty. \tag{6.38}$$

This series is absolutely convergent if the sum of the lengths converges as an ordinary series of real numbers, namely

$$\sum_{n=0}^{\infty} \|x_n\| < \infty. \tag{6.39}$$

In a complete metric space such as the Hilbert space H a series of vectors in H which converges absolutely also converges to an element x in H, namely

$$\left\| x - \sum_{n=0}^{N} x_n \right\| \longrightarrow 0, \quad N \longrightarrow \infty. \tag{6.40}$$

Hilbert spaces are complete normed vector spaces and therefore they are Banach spaces (in this case the basic structure is the norm not the inner product as in the case of the Hilbert space). Hilbert spaces are also examples of topological vector spaces (vector spaces which are also topological spaces with a uniform structure).

The closed linear subspaces of a Hilbert space, which play a major role in quantum logic, are metric spaces with the inner product induced by restriction to the subspaces. These subspaces are also complete and hence they are Hilbert spaces in their own right.

A summary of quantum mechanics

The mathematical framework of Hilbert space quantum mechanics consists in the following:
- A pure state $|\psi\rangle$ is a point x in the Hilbert space H, i.e. $x \equiv |\psi\rangle$. More precisely, a pure physical state $|\psi\rangle$ is a normalized vector in H which corresponds to the one-dimensional closed linear subspace $M = \{\alpha|\psi\rangle, \quad \alpha \in \mathbb{C}\}$. A pure physical state $|\psi\rangle$ can also be given by the orthogonal projector P onto M, i.e. $P = |\psi\rangle\langle\psi|$.
- If a physical system can be found in different pure states $|\phi_i\rangle$ with probability amplitudes a_i then the actual pure state of the system is the linear superposition $|\psi\rangle = \sum_i a_i |\phi_i\rangle$. This the superposition principle.
- The probability to find a system represented by a normalized pure state $|\psi\rangle$ in some other normalized pure state $|\phi_i\rangle$ is given by $|\langle\phi_i|\psi\rangle|^2 = |a_i|^2$. This is the Born rule or the statistical interpretation.
- Observables are represented by Hermitian operators O on the Hilbert space H which satisfy $(x, Oy) = (Ox, y)$ for any $x, y \in H$, i.e. $O^\dagger = O$.
- The results of measurements of an arbitrary observable in an arbitrary state $|\psi\rangle \in H$ are given by the eigenvalues λ_i of the Hermitian operator O associated with that observable. The eigenvalue λ_i is obtained as an outcome of the measurement with a probability $|\langle\psi|\lambda_i\rangle|^2$, where $|\lambda_i\rangle$ is the eigenvector associated with the eigenvalue. The state of the system $|\psi\rangle$ after the measurement of the observable

O will collapse to the eigenvector $|\lambda_i\rangle$. This is the collapse or reduction postulate. This is postulate is the most problematic aspect of quantum mechanics.
- The expectation value of the observable O is given by $\langle O \rangle = \sum_i \lambda_i |\langle \psi | \lambda_i \rangle|^2$. This is equivalent to the statistical interpretation or the Born rule.
- The time evolution of the pure state $|\psi\rangle$ of the system is governed by Schrödinger's equation $i\hbar \frac{\partial}{\partial t} |\psi\rangle t = H |\psi\rangle$.
- The state of the system is described by a density matrix ρ in the non-pure, i.e. mixed state case.

6.2.2 Experimental propositions

When we go to quantum theory the phase space is replaced with a complex separable Hilbert space **H**. States (pure states) are given now by vectors in **H** (more precisely normalized vectors corresponding to rays) while physical observables are given by Hermitian operators on **H** which are in general incompatible operators, i.e. they do not commute under the point-wise multiplication of operators. Measurement of these physical observables will yield the eigenvalues of the corresponding Hermitian operators. The fact that these operators are non-commuting means that the corresponding eigenvalues do not exist as measurement values prior to (and independent of) the measurement process in stark contrast with classical mechanics.

In classical mechanics experimental propositions are represented by subsets of the phase space Σ whereas the corresponding negated propositions are represented by the complements of these subsets with respect to the phase space. In the Hilbert space subsets may contain, by the principle of linear superposition, both properties and their negations and as a consequence subsets in the Hilbert space are not good representations of experimental propositions.

According to Birkhoff and von Neumann experimental propositions P in quantum mechanics are given by the closed linear subspaces of the Hilbert space **H** which are in one-to-one correspondence with projection operators or projectors. This can be seen as follows. Let M be some closed linear subspace of H and let $M^\perp$ be its complement, i.e.

$$M^\perp = \{|\chi\rangle \in H : \forall |\phi\rangle \in M (\langle \chi, \phi \rangle = 0)\}. \tag{6.41}$$

Any vector $|\psi\rangle$ in H can be decomposed in a unique way as the sum of two orthogonal vectors $|\phi\rangle \in M$ and $|\chi\rangle \in M^\perp$, namely

$$|\psi\rangle = |\phi\rangle + |\chi\rangle. \tag{6.42}$$

The projector P corresponding to the closed linear subspace M is given by

$$P|\psi\rangle = |\phi\rangle, \quad P^\dagger = P, \quad P^2 = P. \tag{6.43}$$

Conversely, every projector P is associated with a closed linear subspace M of H.

This one-to-one map (isomorphism) allows us to translate the lattice structure of the set **C**(H) of all closed linear subspaces of H into the algebraic structure of the set **P**(H) of all projection operators [9].

According to Birkhoff and von Neumann experimental propositions are precisely the projectors P (and hence our use of the same symbol to denote both). These experimental propositions are of the typical form:
- P: The physical system S has a property λ, i.e. if an observable $\mathcal{O}$ is measured, then we will obtain as outcome the value λ.

After measuring the observable $\mathcal{O}$ we will either obtain the value λ (and therefore the proposition P is true) or we will not (in which case the proposition P is false). This is therefore a true–false proposition which can be represented by a projector in a natural way as follows.

The observable $\mathcal{O}$ is a Hermitian operator in the Hilbert space H which we will assume to be finite dimensional (dimension n). Every Hermitian operator $\mathcal{O}$ in a finite-dimensional Hilbert space H admits a unique spectral decomposition consisting of $r \leqslant n$ real eigenvalues $\lambda_1, \ldots, \lambda_r$ with eigenspaces given by the orthogonal projectors $P_1, \ldots, P_r$. Thus, we have

$$\mathcal{O} = \sum_{i=1}^{r} \lambda_i P_i, \quad \sum_{i=1}^{r} P_i = 1. \tag{6.44}$$

In other words, the eigenvalues λ_i correspond precisely to the properties λ of the system S obtained as measurement outcomes whereas the corresponding projectors correspond precisely to the true–false propositions P. It is quite remarkable that any measurement is decomposable into these elementary logical true–false propositions [9].

The eigenvalues of any projector P_i are either 0 or 1 and thus the expectation value $\langle u|P_i|u\rangle$ of this projector in a normalized state vector $|u\rangle$ describing the system S will lie clearly in the interval [0, 1]. This expectation value can therefore be interpreted as the probability that a measurement of the observable $\mathcal{O}$ will lead the eigenvalue λ_i, i.e. the probability that the proposition 'P: The physical system S has a property λ' is true. In fact, the proposition P is true if and only if $P_i|u\rangle = |u\rangle$, i.e. the state $|u\rangle$ is found in the linear subspace M_i associated with the projector M_i [10].

The isomorphism between closed linear subspaces M of H and projectors P on one hand and also the isomorphism between projectors and propositions on the other hand permits us to use these terms (propositions, projectors, closed linear subspaces) interchangeably.

In summary, events in classical mechanics and classical logic are subsets of the phase space Σ and pure states semantically decide the truth value of experimental propositions. Hence a given point x (pure state) in phase space either belongs to a given subset X of Σ (x decides that the experimental proposition P represented by the classical event X is true) or it does not belong to X (x decides that P is false). Thus classical logic satisfies the principle of excluded middle.

In quantum mechanics the situation is drastically different since the underlying quantum logic is effectively probabilistic. Events in this case are closed linear subspaces M of the Hilbert space H represented by projectors P. A pure state $|\psi\rangle$ in H decides the truth value of an experimental proposition P with the following probabilities [6]:

- The proposition P is true: the state $|\psi\rangle$ assigns to the proposition P a probability $\mathcal{P} = 1$.
- The proposition P is false: the state $|\psi\rangle$ assigns to the proposition P a probability $\mathcal{P} = 0$.
- The proposition P is semantically undecided or undetermined: the state $|\psi\rangle$ assigns to the proposition P a probability $\mathcal{P} \neq 0, 1$.

Let us now assume that the state $|\psi\rangle$ is actually a linear superposition of several states $|\psi_i\rangle$ with coefficients c_i, namely

$$|\psi\rangle = \sum_i c_i |\psi_i\rangle. \tag{6.45}$$

The state $|\psi\rangle$ will then satisfy with a probability $|c_i|^2$ all the experimental propositions satisfied with certainty by the state $|\psi_i\rangle$, i.e. with a probability $\mathcal{P}_i = 1$. Hence if the states $|\psi_i\rangle$ satisfy an experimental proposition P with certainty then the linear superposition $|\psi\rangle$ will also satisfy the experimental proposition P with certainty provided

$$\sum_i |c_i|^2 = 1. \tag{6.46}$$

Hence subsets of the Hilbert space H corresponding to the propositions P are subsets which are closed under linear combinations, i.e. they are the closed linear subspaces associated with projectors [6, 8, 9]. This shows in a dramatic way the fundamental role played by the superposition principle in quantum mechanics.

6.2.3 The logic of projectors

Any closed linear subspace M of the Hilbert space corresponds therefore to an experimental proposition and they are both represented by a projector. Both the proposition and the projector are denoted by the same symbol P. The proposition is a simple true–false statement which in English is of the typical form:
- P: The physical system S has a property λ, i.e. if we measure an observable $\mathcal{O}$ we will obtain the value λ as an outcome. The value λ is the eigenvalue of $\mathcal{O}$ with eigenspace M defined by the projector P.

Quantum events are therefore closed linear subspaces M of H represented by projectors P. On the set $\mathbf{C}(H)$ of quantum events (which is isomorphic to the set $\mathbf{P}(H)$ of all projectors on H) an algebraic structure (called a lattice) can be defined as follows [9, 11]:
- The logical **AND** or conjunction operation $\wedge$ (also called meet). Let P and Q be two experimental propositions associated with the closed linear subspaces M_P and M_Q, the statement $P \wedge Q$ is also a valid experimental proposition associated with the set-theoretical intersection $M_P \cap M_Q$ of the two subspaces M_P and M_Q which is also a closed linear subspace of H. We have then

$$M_{P \wedge Q} = M_P \cap M_Q = \{x: x \in M_P, x \in M_Q\}. \tag{6.47}$$

Thus the conjunction operation is the same as intersection in classical logic.

For two commuting projectors P and Q the closed linear subspace $M_{P \wedge Q}$ associated with the proposition $P \wedge Q$ corresponds to the projector $P.Q$.

- The logical **OR** or inclusive disjunction operation $\vee$ (also called join). Let P and Q two experimental propositions associated with the closed linear subspaces M_P and M_Q, the statement $P \vee Q$ is also a valid experimental proposition associated not with the set-theoretical union $M_P \cup M_Q$ of the two subspaces M_P and M_Q, but it is associated with the direct sum $M_P \oplus M_Q$ of the two closed linear subspaces M_P and M_Q. Indeed the direct sum $M_P \oplus M_Q$ is the smallest closed linear subspace of H which includes both M_P and M_Q and their union $M_P \cup M_Q$. We have then

$$M_{P \vee Q} = M_P \oplus M_Q = \{x: x = \alpha y + \beta z, y \in M_P, y \in M_Q, \alpha, \beta \in \mathbb{C}\}. \quad (6.48)$$

Thus the disjunction operation in quantum logic is very different from the disjunction operation defined in classical logic.

For two commuting projectors P and Q the closed linear subspace $M_{P \vee Q}$ associated with the proposition $P \vee Q$ corresponds to the projector $P + Q - P.Q$.

- The logical **NOT** operation $\neg$. Let P be an experimental proposition associated with the closed linear subspace M_P, the experimental proposition $\neg P$ is the negation of P if it is associated with the closed linear subspace $M_P^\perp$ of H where $M_P^\perp$ is the orthogonal complement (orthocomplement) of the subspace M_P. We have then

$$M_{\neg P} = M_P^\perp = \{x: \langle x, y \rangle = 0, y \in M_P\}. \quad (6.49)$$

We have clearly the correct behavior:
- If P is true then $\neg P$ is false: The state $|\psi\rangle$ assigns to the proposition P a probability $\mathcal{P} = 1$ and to the proposition $\neg P$ a probability $\mathcal{P}^\perp = 0$.
- If P is false then $\neg P$ is true: The state $|\psi\rangle$ assigns to the proposition P a probability $\mathcal{P} = 0$ and to the proposition $\neg P$ a probability $\mathcal{P}^\perp = 1$.

Obviously, the projector associated with the closed linear subspace $M_{\neg P}$ is $1 - P$.

- The ordering relation $\leqslant$: This is given here by the set-theoretical inclusion $\subseteq$ which acts as the material conditional $\longrightarrow$ between experimental propositions. Thus, if P and Q are two experimental propositions associated with the closed linear subspaces M_P and M_Q, then $P \longrightarrow Q$ if and only if $M_P \subseteq M_Q$. We have then

$$P \longrightarrow Q \iff M_P \subseteq M_Q \iff M_P \cap M_Q = M_P. \quad (6.50)$$

For two commuting projectors this is equivalent to $P.Q = P$.

- A trivial proposition (a tautology) is a statement which is always true. This is denoted by **1** and it is the identity with respect to the conjunction operation $\wedge = \cap$. This corresponds to the entire Hilbert space, i.e.

$$M_1 = H. \quad (6.51)$$

- An impossible proposition (an absurdity) is a statement which is always false. This is denoted by **0** and it is the identity with respect to the disjunction operation $\vee = \cup$. This corresponds to the empty set, i.e.

$$M_0 = \varnothing. \tag{6.52}$$

The non-Boolean algebra or lattice ($\mathbf{C}(H)$, $\subseteq$) which encodes the quantum logic of projectors is given by the structure

$$(\mathbf{C}(H), \subseteq, \cap, \oplus, \perp, H, \varnothing). \tag{6.53}$$

This is an orthocomplemented modular lattice which is non-distributive and therefore non-Boolean. This lattice is called a Hilbert lattice and is denoted by $\mathbf{L}(H)$. We have the following properties:

- The conjunction (meet) $\wedge = \cap$ and the disjunction (join) $\vee = \oplus$ operations are commutative. For any x, y in $\mathbf{C}(H)$ we have

$$x \wedge y = y \wedge x, \quad x \vee y = y \vee x. \tag{6.54}$$

- The conjunction (meet) $\wedge = \cap$ and the disjunction (join) $\vee = \oplus$ operations are associative. For any x, y, z in $\mathbf{C}(H)$ we have

$$x \wedge (y \wedge z) = (x \wedge y) \wedge z, \quad x \vee (y \vee z) = (x \vee y) \vee z. \tag{6.55}$$

- The conjunction (meet) $\wedge = \cap$ and the disjunction (join) $\vee = \oplus$ operations satisfy the absorption law. For any x, y in $\mathbf{C}(H)$ we have

$$x \wedge (x \vee y) = x, \quad x \vee (x \wedge y) = x. \tag{6.56}$$

- The lattice $\mathbf{L}(H)$ is a bounded lattice with a maximum given by the Hilbert space H (which acts as the identity **1** with respect to the meet operation $\wedge$) and a minimum given by the empty set $\varnothing$ (which acts as the identity **0** with respect to the join operation $\vee$).
- This lattice $\mathbf{L}(H)$ is an orthocomplemented lattice or ortholattice since for any element $x \in \mathbf{C}(H)$ an orthocomplement $x^\perp \in \mathbf{C}(H)$ exists given by the orthogonal complement. The relation of orthogonal complement $\perp$ satisfies the following

$$\begin{aligned} &x^\perp \vee x = \mathbf{1}, \quad x^\perp \wedge x = \mathbf{0} \\ &(x^\perp)^\perp = x \\ &x \leqslant y \iff y^\perp \leqslant x^\perp. \end{aligned} \tag{6.57}$$

- *The modular law*: The ortholattice $\mathbf{L}(H)$ is called modular since it satisfies the so-called modular law which states that for any x, y, z in $\mathbf{C}(H)$ such that $x \leqslant z$ (or equivalently $x \subseteq z$) we have $x \vee (y \wedge z) = (x \vee y) \wedge z$. We have then

$$x \leqslant z \Rightarrow x \vee (y \wedge z) = (x \vee y) \wedge z. \tag{6.58}$$

This is a weaker condition than the distributive law which actually holds only if H is a finite-dimensional Hilbert space. It is equivalent to

$$x \leqslant y \Rightarrow x \vee (y \wedge z) = y \vee (x \wedge z). \tag{6.59}$$

If we set $y = x^\perp$ in the modular law (6.66) we obtain a weaker condition (called the orthomodular condition) which holds in any Hilbert space given by [12]

$$x \leqslant z \Rightarrow x \vee (x^\perp \wedge z) = z. \tag{6.60}$$

In this case the ortholattice $L(H)$ is called orthomodular.

- *The distributive law*: The modular law is actually a weakening of the distributive law which can also be given by the relations [6]

$$(x \wedge y) \vee (x \wedge z) \subseteq x \wedge (y \vee z), \quad \forall x, y, z \in \mathbf{C}(H). \tag{6.61}$$

$$x \vee (y \wedge z) \subseteq (x \vee y) \wedge (x \vee z), \quad \forall x, y, z \in \mathbf{C}(H). \tag{6.62}$$

In a distributive lattice the ordering relation $\leqslant = \subseteq$ in the above two equations is replaced with an equality $=$.

The failure of the distributive law in the ortholattice $L(H)$ is related to the fact that the disjunction (join) operation $\vee$ is not the ordinary set-theoretical union $\cup$ but it is actually the direct sum (linear span) $\oplus$ since $x \oplus y$ gives the smallest closed linear subspace of H which contains the union $x \cup y$ of the two closed linear subspaces x and y. Hence a quantum disjunction $x \vee y$ can be true even if neither x nor y are true. Indeed, a pure state $|\psi\rangle$ may belong to $x \oplus y$ (due to linear superposition) even if $|\psi\rangle$ belongs neither to x nor to y [6].

6.3 Hilbert lattices

In this section we will follow mainly the book [9]. A more technical reference on orthomodular lattices is [13].

6.3.1 Lattice theory

- *Posets*: A partially ordered set (a poset) is a set S with ordering relations $\leqslant$ which satisfies: (i) reflexivity, i.e. $x \leqslant x$, $\forall x \in S$, (ii) transitivity, i.e. $x \leqslant y$ and $y \leqslant z$ then $x \leqslant z$, $\forall x, y, z \in S$, and (iii) identitivity, i.e. $x \leqslant y$ and $y \leqslant x$ then $x = y$.

    We say that x covers y if $y \leqslant x$ and there is no $z \in S$ such that $y \leqslant z \leqslant x$, i.e. x is the next element above y.

- *Hasse diagram*: A finite poset $(S, \leqslant)$ can be represented by the so-called Hasse diagram. Every element x in S is represented higher than all elements y in S

such that $y \leqslant x$ with segments drawn then from y to x. All elements are represented by points.
- *Atoms*: In a partially ordered set with a least element **0** and a greatest element **1** atoms are elements which cover **0** whereas co-atoms are elements which are covered by **1**.

  A partially ordered set is called atomic if every element $x \neq \mathbf{0}$ is greater than or equal to an atom.
- *Lattices* $(S, \leqslant)$: A lattice L is a partially ordered set $(S, \leqslant)$ in which every pair $(x, y) \in S \times S$ has:
  - A meet (greatest lower bound) $\inf(x, y)$ such that $\inf(x, y) \leqslant x$, $\inf(x, y) \leqslant y$, if $z \leqslant x$ and $z \leqslant y$ then $z \leqslant \inf(x, y)$.
  - A join (least upper bound) $\sup(x, y)$ such that $x \leqslant \sup(x, y)$, $y \leqslant \sup(x, y)$, if $x \leqslant z$ and $y \leqslant z$ then $\sup(x, y) \leqslant z$.
- *Lattices* $(S, \wedge, \vee)$: A lattice L is an algebraic structure $(S, \wedge, \vee)$ with the two operations $\wedge$ and $\vee$ satisfying:
  - *Commutativity*: $x \wedge y = y \wedge x$ and $x \vee y = y \vee x$ for any x, y in S.
  - *Associativity*: $x \wedge (y \wedge z) = (x \wedge y) \wedge z$ and $x \vee (y \vee z) = (x \vee y) \vee z$ for any x, y, z in S.
  - *Absorption law*: $x \wedge (x \vee y) = x$, $x \vee (x \wedge y) = x$ for any x, y in S.
  - *Idempotence*: $x \wedge x = x$ and $x \vee x = x$ for any x in S.

  The two structures $(S, \leqslant)$ and $(S, \wedge, \vee)$ are equivalent with the identifications:

$$\begin{aligned} x \leqslant y &\iff x \wedge y = x \text{ or } x \vee y = y \\ \inf(x, y) &= x \wedge y \\ \sup(x, y) &= x \vee y. \end{aligned} \tag{6.63}$$

- *Identity elements*: The upper bound **1** and the lower bound **0** of a lattice satisfy

$$\begin{aligned} \mathbf{0} \wedge x &= \mathbf{0}, & \mathbf{0} \vee x &= x \\ \mathbf{1} \wedge x &= x, & \mathbf{1} \vee x &= \mathbf{1}. \end{aligned} \tag{6.64}$$

Every finite lattice (one with finite number of elements) contains an upper bound **1** and a lower bound **0**.
- *Orthocomplemented lattices*: The orthogonal complement or orthocomplement $\neg x$ of an element $x \in S$ is an element in S satisfying $\neg(\neg x) = x$, $\neg x \wedge x = \mathbf{0}$, $\neg x \vee x = \mathbf{1}$, and $x \leqslant y \Rightarrow \neg y \leqslant \neg x$.

  An orthocomplemented lattice L is a lattice in which every element $x \in S$ has an orthocomplement $\neg x$ which is also in S.
- *Subalgebras/sublattices*: A subalgebra of an orthocomplemented lattice $L = (S, \wedge, \vee)$ is a subset of S which is closed under the operations $\wedge, \vee, \neg$ and which contains the identity elements **1**, **0**.

  A sublattice of an orthocomplemented lattice $L = (S, \wedge, \vee)$ is a subset of S which is only closed under the operations $\wedge, \vee$ but not under $\neg$.

- *Distributive lattices*: A lattice L is called a distributive lattice if
$$x \wedge (y \vee z) = (x \wedge y) \vee (x \wedge z),$$
$$x \vee (y \wedge z) = (x \vee y) \wedge (x \vee z), \quad \forall x, y, z \in S. \tag{6.65}$$

- *Boolean lattices*: An orthocomplemented lattice which is also distributive is called a Boolean lattice.
- *Modular lattices*: If we assume that $x \leqslant z$, i.e. $z = x \vee z$, in the distributive law $x \vee (y \wedge z) = (x \vee y) \wedge (x \vee z)$ we obtain immediately the condition
$$x \vee (y \wedge z) = (x \vee y) \wedge z, \quad \forall x, y, z \in S. \tag{6.66}$$
This is called the modular condition. Therefore any distributive lattice is a modular lattice but the converse is not true. The modular condition is clearly a weaker form of the distributive law.

   A lattice L is called a modular lattice if for every $x \leqslant z$ the condition (6.66) holds.
- *Orthomodular lattices*: If we choose $y = \neg x$ in the condition (6.66) we obtain
$$x \vee (\neg x \wedge z) = z, \quad \forall x, z \in S. \tag{6.67}$$
This is called the orthomodular condition. Therefore any modular lattice is an orthomodular lattice but again the converse is not true.

   As it turns out, when we deal with quantum logic, the modular condition can only hold in finite-dimensional Hilbert spaces. In general (infinite-dimensional Hilbert spaces) only the orthomodular condition is satisfied. The orthomodular condition is a weaker form of the modular condition.

   A lattice L is called an orthomodular lattice if for every $x \leqslant z$ the condition (6.67) holds.
- *Blocks*: As it turns out, orthomodular lattices can be decomposed into Boolean subalgebras. The maximal Boolean subalgebra (containing maximum number of atoms) of an orthomodular lattice is called a block.
- *Pasting*: By starting from a set of Boolean lattices $\{L_i\}$ we can construct an orthomodular lattice L by taking the union of the Boolean lattices L_i, i.e. $L = \cup L_i$, provided that for all $i \neq j$ we have:
  - $L_i \not\subset L_j$.
  - $L_i \cap L_j$ is an orthomodular Boolean sublattice where the partial orderings and orthocomplementations of L_i and L_j coincide on the intersection $L_i \cap L_j$.

   The set $L = \cup L_i$ is called the pasting of $\{L_i\}$. The partial ordering and orthocomplementation on L are defined by
$$x \leqslant |_L y \iff x \leqslant |_{L_i} y \quad \text{for some } L_i, \tag{6.68}$$
$$y = \neg x|_L \iff y = \neg x|_{L_i} \quad \text{for some } L_i. \tag{6.69}$$

The problem of pasting, i.e. the construction of orthomodular lattices from Boolean algebras, is the inverse problem to the block-decomposition problem, i.e. finding the maximal Boolean subalgebra of an orthomodular lattice.

As it turns out, every orthomodular lattice is a pasting of its blocks while every Hilbert lattice is an irreducible pasting of its blocks.

6.3.2 Hilbert lattices

Let H be a separable Hilbert space. The Hilbert lattice L is an algebraic structure based on the set $\mathbf{C}(H)$ of all closed linear subspaces of H which is isomorphic to the set $\mathbf{P}(H)$ of all projectors on H. Experimental propositions (quantum events) are elements of $\mathbf{C}(H)$ or equivalently elements of $\mathbf{P}(H)$. The partial ordering on the set $\mathbf{C}(H)$ is the set-theoretical inclusion $\subseteq$ and it plays the role of implication (material conditional $\longrightarrow$). The meet or **AND** operation is the set-theoretical intersection $\cap$ while the join or **OR** operation is the direct sum (closed linear span) $\oplus$ (and *not* the set-theoretical union $\cup$ which is the single most important difference with classical logic and Boolean algebras). The negation $\neg P$ of an experimental proposition P corresponds to its orthogonal closed linear subspace given by the orthogonal projector $P^\perp$, i.e. $\neg P = P^\perp$. The Hilbert lattice L is therefore given by

$$L = (\mathbf{C}(H), \subseteq) = (\mathbf{C}(H), \cap, \oplus, \perp, \mathbf{1}, \mathbf{0}). \tag{6.70}$$

Obviously, the trivial proposition $\mathbf{1}$ is H while the impossible proposition $\mathbf{0}$ is $\emptyset$. The trivial proposition $\mathbf{1}$ is therefore a tautology.

Other tautologies of the above quantum logic are the fundamental axioms of the lattice structure $L = (\mathbf{C}(H), \leq)$ given by: idempotence, commutativity, associativity and absorption. These tautologies also exist in the classical logic. However, the quantum logic $L = (\mathbf{C}(H), \leq)$ is intrinsically non-Boolean, in marked contrast with the Boolean structure of the classical logic, since the distributive law fails in the Hilbert lattice L.

The Hilbert lattice $L = (\mathbf{C}(H), \leq)$ is actually a modular lattice in the case of finite-dimensional Hilbert spaces whereas it is an orthomodular lattice in the general case. The orthomodular condition is a weaker form of the modular condition while the modular condition is a weaker form of the distributive law.

In a finite-dimensional Hilbert space the modular conditions read

$$P_1 \vee (P_2 \wedge P_3) = (P_1 \vee P_2) \wedge P_3, \quad \text{for all } P_1 \longrightarrow P_3. \tag{6.71}$$

$$P_1 \wedge (P_2 \vee P_3) = (P_1 \wedge P_2) \vee P_3, \quad \text{for all } P_3 \longrightarrow P_1. \tag{6.72}$$

In an arbitrary Hilbert space (which is the generic case) the orthomodular condition reads

$$P \vee (P^\perp \wedge Q) = Q, \quad \text{for all } P \longrightarrow Q. \tag{6.73}$$

6.3.3 Hilbert lattice as pasting of blocks

We have already seen that the problem of pasting, i.e. the construction of orthomodular lattices starting from Boolean algebras, is the inverse problem to the block-decomposition problem, i.e. finding the maximal Boolean subalgebras of a given orthomodular lattice. In other words an orthomodular lattice, perhaps such as the Hilbert lattice, can be viewed as a pasting of blocks given by Boolean subalgebras.

Since a block is a maximal Boolean subalgebra it should naturally correspond to a maximal number of compatible and therefore comeasurable observables. The Hilbert lattice contains of course many distinct blocks which are non-comeasurable corresponding to incompatible collections of observables. Every block is defined locally, i.e. locally measurable, but it cannot be defined globally independently of the simultaneous measurement of the other blocks. In other words, at any one time only one block can really be measured in any collection of non-comeasurable blocks.

But is there really a difference between an observable that is actually measured (the measured block) and one which is only obtained by counterfactual argument (the other non-comeasurable blocks)?

Realists such as Einstein, Podolsky and Rosen [14] and those who subscribe to hidden variables seem to answer this question with a resounding 'no'. Thus, there is no fundamental difference between an observable that is measured and one which is only obtained by counterfactual arguments, and hence we can deal consistently with non-comeasurable blocks simultaneously.

The other extreme is the statement that 'unperformed experiments have no results' [15, 16] which corresponds to answering the above question with a resounding 'yes'. Thus, there is a real fundamental difference between an observable that is measured and one which is only obtained by counterfactual arguments, and hence non-comeasurable observables make no physical sense. Indeed, Bell's theorem [17] seems to suggest that the result of measurement of one block are context-dependent of the measurement of the other blocks, i.e. counterfactuality is not free but is bounded by contextuality (Kochen–Specker theorem [18]).

Here we will follow [9], who takes the middle ground that non-comeasurable observables make perfect mathematical sense, which allows us to understand conceivable physical phenomena better. A quantum non-Boolean logic is then constructed by pasting together Boolean subalgebras (blocks) L_i as follows [9]:
- The trivial propositions of all blocks are identified.
- The impossible propositions of all blocks are identified.
- Identical propositions in different blocks are identified.
- The logic and algebraic structures in all blocks remain the same.

The lattice obtained is denoted by $L_1 \oplus L_2 \oplus \ldots$ and it provides a global representation of the local universes L_i.

6.3.4 Spin one-half system

For a finite-dimensional Hilbert space the quantum propositional logic corresponding to a maximal block is 2^n-dimensional, where n is the dimension of the Hilbert space.

The Hilbert space in the case of a system of a spin one-half particle is two-dimensional. For simplicity we will assume that this Hilbert space is real-valued, i.e. $H = \mathbb{R}^2$. We will consider the measurements of the spin along the x-axis, i.e. the measurements of the compatible observables $\vec{S}^2$ and S_x. Obviously, the results of measurements are either spin up $|+\rangle$ corresponding to $\vec{S}^2 = 3\hbar^2/4$, $S_x = \hbar/2$ or spin down $|-\rangle$ corresponding to $\vec{S}^2 = 3\hbar^2/4$, $S_x = -\hbar/2$. The comeasurable block is a maximal Boolean subalgebra L_x (of dimension 2^2) consisting of the following propositions:

- P_+: The system is in the state $|+\rangle$. This proposition corresponds to the one-dimensional closed linear subspace of H spanned by the vector $|+\rangle = (1, 0)$. This subspace is associated with the projector $P_+ = |+\rangle\langle+|$.
- P_-: The system is in the state $|-\rangle$. This proposition corresponds to the one-dimensional closed linear subspace of H spanned by the vector $|-\rangle = (0, 1)$. This subspace is associated with the projector $P_- = |-\rangle\langle-|$.
- The trivial proposition **1** or 'the system is in some spin state' is always true and is identified with the entire Hilbert space H.
- The impossible proposition **0** or 'the system is in no state at all' is always false and is identified with the zero-dimensional closed linear subspace $\emptyset$. In this case the system can be thought of as occupying the zero vector $(0, 0)$.

The two propositions P_+ and P_- are the atoms of this logical system since they cover the impossible proposition **0** and they are also co-atoms since they are covered by the trivial proposition **1**. They obviously correspond to the orthogonal lines passing through the origin in $H = \mathbb{R}^2$. Any other line passing through the origin (closed linear subspaces of H) spanned by a non-zero state vector corresponds to the proposition that the system occupies that state. The negation of P_+ and P_- are given by P_- and P_+, respectively, since the complement of the corresponding lines are the lines in H orthogonal to them. The complement of the entire Hilbert space H is the empty set $\emptyset$ and the complement of the empty set $\emptyset$ is the entire Hilbert space H and thus the negation of **1** is **0** and the negation of **0** is **1**. Hence, L_x is a complemented lattice, where the complement corresponds to the orthogonal complement in the Hilbert space H, which is also distributive, i.e. it is a Boolean algebra. The associated Hasse diagram is shown in figure 6.1.

In summary, the comeasurable block associated with the measurements of the spin along the x-axis is given by the Boolean subalgebra

$$L_x = \{P_+, P_-, \mathbf{1}, \mathbf{0}\}. \tag{6.74}$$

Similarly, the comeasurable block associated with the measurements of the spin along the z-axis is given by the Boolean subalgebra (with $\bar{x} \equiv z$)

$$L_{\bar{x}} = \{\bar{P}_+, \bar{P}_-, \bar{\mathbf{1}}, \bar{\mathbf{0}}\}. \tag{6.75}$$

The two blocks L_x and $L_{\bar{x}}$ are clearly non-comeasurable. The pasting of the two subalgebras L_x and $L_{\bar{x}}$ is a quantum non-distributive non-Boolean logic given by the orthocomplemented modular lattice

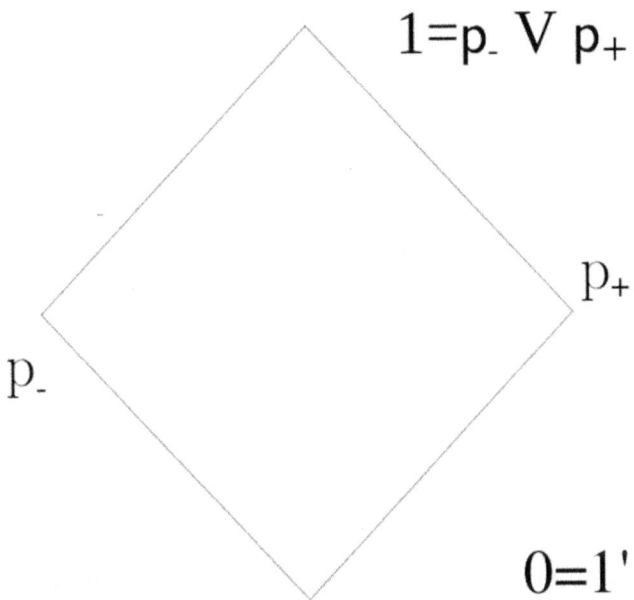

Figure 6.1. Hasse diagram of the Boolean algebra L_x for spin one-half.

$$L_x \oplus L_{\bar{x}} = MO_2. \tag{6.76}$$

In this lattice we make the identification

$$\mathbf{1} = \bar{\mathbf{1}}, \quad \mathbf{0} = \bar{\mathbf{0}}. \tag{6.77}$$

This pasting of the two blocks L_x and $L_{\bar{x}}$ at the level of the tautologies and absurdities gives what is called a horizontal sum, i.e. MO_2 is a horizontal sum of L_x and $L_{\bar{x}}$. The associated Hasse diagram is shown in figure 6.2. The non-distributive character of this lattice can be shown as follows [9, 11]. On the one hand, we have

$$P_- \vee \left(\bar{P}_- \wedge \bar{P}_-^\perp \right) = P_- \vee \mathbf{0} = P_-. \tag{6.78}$$

On the other hand, if we assume the distributive law, we have

$$P_- \vee \left(\bar{P}_- \wedge \bar{P}_-^\perp \right) = \left(P_- \vee \bar{P}_- \right) \wedge \left(P_- \vee \bar{P}_-^\perp \right)$$
$$= \mathbf{1} \wedge \mathbf{1}$$
$$= \mathbf{1}. \tag{6.79}$$

Thus, we obtain a contradiction. In the second line of (6.79) we have used the fact that $a \vee b$ is the least upper bound of a and b and hence $a \vee b$ is the highest element in the Hasse diagram which is connected to a and b and therefore $P_- \vee \bar{P}_- = \mathbf{1}$ and similarly $P_- \vee \bar{P}_-^\perp = \mathbf{1}$.

Similarly, we can show that

$$P_- \wedge \left(\bar{P}_- \vee \bar{P}_-^\perp \right) = P_- \wedge \mathbf{1} = P_-. \tag{6.80}$$

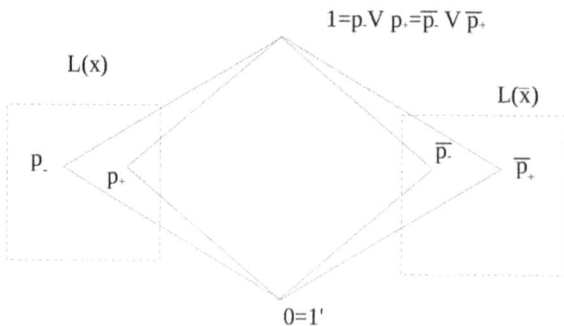

Figure 6.2. Hasse diagram of the Boolean algebra $L_x \oplus L_{\bar{x}}$.

$$P_- \wedge \left(\bar{P}_- \vee \bar{P}_-^\perp\right) = \left(P_- \wedge \bar{P}_-\right) \vee \left(P_- \wedge \bar{P}_-^\perp\right)$$
$$= 0 \vee 0 \qquad (6.81)$$
$$= 0.$$

Again this is a contradiction. In the second line of (6.81) we have used the fact that $a \wedge b$ is the greatest lower bound of a and b and hence $a \wedge b$ is the lowest element in the Hasse diagram which is connected to a and b and therefore $P_- \wedge \bar{P}_- = \mathbf{0}$ and similarly $P_- \wedge \bar{P}_-^\perp = \mathbf{0}$.

Generalization of the orthocomplemented modular lattice MO_2 is given by the orthocomplemented modular lattices MO_n which are obtained by pasting together (taking the horizontal sum of) n blocks L_{x_i} (Boolean subalgebras) corresponding to the results of measurements of the spin along n different axes x_i, namely

$$MO_n = \oplus_{i=1}^n L_{x_i}. \qquad (6.82)$$

These are the finite subalgebras of the two-dimensional Hilbert space H [19]. The Hilbert space H is effectively the horizontal sum MO_∞ of an infinite number of Boolean blocks (infinitely many as well as continuously many). Of course at any given time we can only measure one single block in the Hilbert space H while the other blocks can only be obtained by counterfactual arguments.

6.3.5 Spin one system

The Hilbert space in the case of a system of spin one particle is three-dimensional. For simplicity we will assume that this Hilbert space is real-valued, i.e. $H = \mathbb{R}^3$. We will consider the measurements of the spin along the x-axis, i.e. the measurements of the compatible observables $\vec{S}^2$ and S_x. The results of measurements are either spin up $|+1\rangle$ corresponding to $\vec{S}^2 = 2\hbar^2$, $S_x = \hbar$ or spin down $|-1\rangle$ corresponding to $\vec{S}^2 = 2\hbar^2$, $S_x = -\hbar$ or spin zero $|0\rangle$ corresponding to $\vec{S}^2 = 2\hbar^2$, $S_x = 0$. The comeasurable block is a maximal Boolean subalgebra L_x (of dimension 2^3) consisting of the following propositions:

- P_{+1}: The system is in the state $|+1\rangle$. This proposition corresponds to the one-dimensional closed linear subspace of H spanned by the vector $|+1\rangle = (1, 0, 0)$.

This subspace is associated with the projector $P_{+1} = |+1\rangle\langle+1|$. This subspace is clearly a line in $\mathbb{R}^3$ passing through the origin. In fact any line passing through the origin in $\mathbb{R}^3$ is spanned by a non-zero vector and corresponds to the experimental proposition that the system occupies that state.

- P_{-1}: The system is in the state $|-1\rangle$. This proposition corresponds to the one-dimensional closed linear subspace of H spanned by the vector $|-1\rangle = (0, 0, 1)$. This subspace is associated with the projector $P_{-1} = |-1\rangle\langle-1|$.
- P_0: The system is in the state $|0\rangle$. This proposition corresponds to the one-dimensional closed linear subspace of H spanned by the vector $|0\rangle = (0, 1, 0)$. This subspace is associated with the projector $P_0 = |0\rangle\langle 0|$.
- The trivial proposition **1** or 'the system is in some spin state' is always true and is identified with the entire Hilbert space H.
- The impossible proposition **0** or 'the system is in no state at all' is always false and is identified with the zero-dimensional closed linear subspace $\emptyset$. In this case the system can be thought of as occupying the zero vector $(0, 0, 0)$.
- $P_{+1}^\perp = P_{-1} \vee P_0$: The negation of the proposition P_{+1}, i.e. the proposition that 'the system is not in the state $|+1\rangle$' is the orthogonal complement of the line associated with P_{+1} given by the plane in $\mathbb{R}^3$ passing through the origin which is orthogonal to that line. This plane is spanned by the vectors $|-1\rangle$ and $|0\rangle$ corresponding to the proposition 'the system is either in the state $|-1\rangle$ or in the state $|0\rangle$' given by the disjunction $P_{-1} \vee P_0$.

In three-dimensional Hilbert space experimental propositions correspond therefore to either lines or planes passing through the origin. The complement of a line is the plane orthogonal to it and the complement of a plane is the line orthogonal to it.

- $P_{-1}^\perp = P_{+1} \vee P_0$: The negation of the proposition P_{-1}, i.e. the proposition that 'the system is not in the state $|-1\rangle$' is the orthogonal complement of the line associated with P_{-1} given by the plane in $\mathbb{R}^3$ passing through the origin which is orthogonal to that line. This plane is spanned by the vectors $|+1\rangle$ and $|0\rangle$ corresponding to the proposition 'the system is either in the state $|+1\rangle$ or in the state $|0\rangle$' given by the disjunction $P_{+1} \vee P_0$.
- $P_0^\perp = P_{+1} \vee P_{-1}$: The negation of the proposition P_0, i.e. the proposition that 'the system is not in the state $|0\rangle$' is the orthogonal complement of the line associated with P_0 given by the plane in $\mathbb{R}^3$ passing through the origin which is orthogonal to that line. This plane is spanned by the vectors $|+1\rangle$ and $|-1\rangle$ corresponding to the proposition 'the system is either in the state $|+1\rangle$ or in the state $|-1\rangle$' given by the disjunction $P_{+1} \vee P_{-1}$.

The propositions P_{+1}, P_{-1} and P_0 are the atoms of this logical system since they cover the impossible proposition **0**, whereas $P_{+1}^\perp$, $P_{-1}^\perp$ and $P_0^\perp$ are co-atoms since they are covered by the trivial proposition **1**.

In summary, the comeasurable block associated with the measurements of the spin along the x-axis is given by the Boolean subalgebra

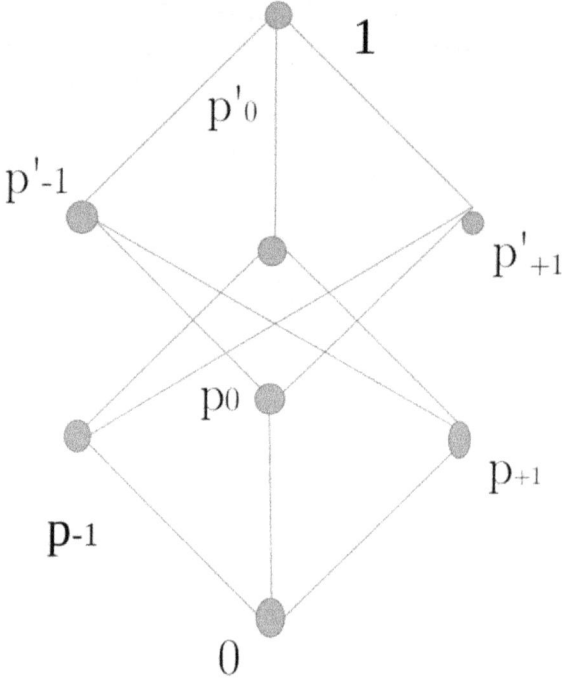

Figure 6.3. Hasse diagram of the Boolean algebra L_x for spin one.

$$L_x = \left\{ P_{+1},\ P_{-1},\ P_0,\ P_{+1}^\perp,\ P_{-1}^\perp,\ P_0^\perp,\ \mathbf{1},\ \mathbf{0} \right\}. \tag{6.83}$$

The Hasse diagram is shown in figure 6.3.

A non-Boolean orthocomplemented modular lattice of the three-dimensional Hilbert space $H = \mathbb{R}^3$ can be obtained by taking the horizontal sum, i.e. pasting together along the tautologies **1** and absurdities **0**, of n non-comeasurable blocks L_{x_i} (Boolean subalgebras of the from (6.83)) corresponding to the results of measurements of the spin along n different axes x_i. However, the spin one system is not an exhausting example of Hilbert lattices of the three-dimensional Hilbert space [9].

6.4 Gleason and the Kochen–Specker theorems

6.4.1 Gleason theorem

In this section we will follow mostly [9, 10].

Hilbert lattices
We consider a separable Hilbert space H and let L be the corresponding Hilbert lattice. We have already established that L is the orthomodular lattice of the set $\mathbf{C}(H)$ of all closed linear subspaces of H which is isomorphic to the set $\mathbf{P}(H)$ of all projectors on H. We will for simplicity write $L = \mathbf{C}(H)$.

Quantum events are experimental propositions and they are elements of $\mathbf{C}(H)$ or equivalently $\mathbf{P}(H)$. A subalgebra of L is a subset of $\mathbf{C}(H)$ which is closed under the logical operations $\wedge = \cap$, $\vee = \oplus$ and $\neg = \perp$ and which contains the identity elements **1** and **0**. A block B of L is a maximal Boolean subalgebra of L ('maximal' means that the subalgebra contains a maximum numbers of atoms which are those elements above **0** directly and 'Boolean' means that the distributive law holds).

The elements of $\mathbf{C}(H)$ and $\mathbf{P}(H)$ will be denoted by M_p and P, respectively, whereas the corresponding propositions will be denoted by p.

Kolmogorov probability theory
The Kolmogorov axioms of probability theory [20] can only be applied to a classical Boolean algebra B of events. A probability measure $\mathcal{P}$ is a function from the set of classical events B into the unit interval, i.e. $\mathcal{P}: B \longrightarrow [0, 1]$ which satisfies (i) $\mathcal{P}(p) \geqslant 0$, $\forall p \in B$, (ii) $\mathcal{P}(B) = 1$, (iii) $\mathcal{P}(p \vee q) = \mathcal{P}(p) \vee \mathcal{P}(q)$ if p and q are mutually disjoint events.

An important example is a maximal Boolean subalgebra B of the Hilbert lattice L with n atoms corresponding to n comeasurable observables. Any positive function $\mathcal{P}$ such that (i) $\mathcal{P}(p_i) \geqslant 0$, $\forall i$, and (ii) $\sum_{i=1}^{n} \mathcal{P}(p_i) = 1$ will satisfy Kolmogorov axioms. Two important examples are:
- The equidistribution $\mathcal{P}(p_i) = 1/n$ (which represents maximum uncertainty or maximum entropy).
- The two-valued dispersion free probability measure $\mathcal{P}(p_i) = \{0, 1\}$ (which represents maximum knowledge and 0 entropy).

An important result is the fact that on Boolean algebras B with n atoms $p_1, \ldots, p_n$ every probability measure is actually a linear combination (a convex sum) of the n two-valued measures $\mathcal{P}_i(p_j) = \delta_{ij}$, $i = 1, \ldots, n$. Both Gleason theorem [21] and the Kochen–Specker theorem [18] state in essence that on a Hilbert lattice, for Hilbert spaces of dimension greater than two, no two-valued probability measure and thus no representation of probability measures as a convex sum exist.

A much more refined notion of classicality is the 'embeddability' of a quantum logic in a classical propositional Boolean structure (more refined than the notion of a 'Boolean' lattice structure). Embeddability means the existence of a two-valued probability measure on the set of quantum events. Indeed, Kochen and Specker showed in [18] that the set of quantum events $\mathbf{C}(H)$ can be embedded into a Boolean algebra B if and only if a two-valued probability measure $\mathcal{P}: \mathbf{C}(H) \longrightarrow \{0, 1\}$ exists such that $\mathcal{P}(p) \neq \mathcal{P}(q)$ if $p \neq q$. We say then that the set of probability measures is 'separating', which is a better characterization of classicality than the 'distributivity' of a Boolean structure.

Gleason theorem
Next we will substitute the Hilbert lattice of quantum events given by the closed subspaces of the Hilbert space for the Boolean algebra of classical events.

The Kolmogorov axioms of probability theory can be applied to the Hilbert lattice L consistently by applying it to each and every Boolean subalgebra B of L as follows. A probability measure $\mathcal{P}$ is a mapping from the set of quantum events or experimental propositions $\mathbf{C}(H)$ into the unit interval, i.e. $\mathcal{P}: \mathbf{C}(H) \longrightarrow [0, 1]$ which satisfies, in each Boolean subalgebra or block B corresponding to comeasurable propositions or events, the following quasi-classical conditions:

- The probability of each proposition p associated with the projection operator P on the subspace M_p is positive, i.e.

$$\mathcal{P}(M_p) \geqslant 0, \quad \forall\, M_p \in \mathbf{C}(H). \tag{6.84}$$

- The sum of all probabilities is one (the Hilbert space corresponds to the tautology **1**), namely

$$\mathcal{P}(H) = 1. \tag{6.85}$$

- The probability measure is countably additive, i.e. for mutually disjoint events p and q represented by orthogonal projection operators P and Q, respectively, we have

$$M_{p \vee q} = M_p \oplus M_q \tag{6.86}$$

and

$$\mathcal{P}(M_{p \vee q}) = \mathcal{P}(M_p) + \mathcal{P}(M_q). \tag{6.87}$$

The propositions p and q belong to the same Boolean subalgebra and therefore they correspond to comeasurable events associated with compatible observables which are represented by commuting self-adjoint operators (projectors) on the Hilbert space.

- However, a projector can belong to different blocks or Boolean algebras representing the different contexts of the measurement of the corresponding proposition (equivalently the blocks represent the different perspectives of the various observers). Thus, at the intersection of two Boolean subalgebras B_1 and B_2 pasted together we must also impose the non-contextuality requirement [22]

$$\mathcal{P}_{B_1}(M_p) = \mathcal{P}_{B_2}(M_p), \quad M_p \in B_1 \cap B_2. \tag{6.88}$$

A probability measure on the subspaces of the Hilbert space satisfying the above requirements can be constructed as follows. Let $|\phi\rangle$ be an arbitrary normalized vector in the Hilbert space H and let P_ϕ be the corresponding one-dimensional projector, i.e. $P_\phi = |\phi\rangle\langle\phi|$. The probability of the projector $P = |\psi\rangle\langle\psi|$ is set to be given by the Born's rule

$$\mathcal{P}_\phi(M_p) = \mathrm{tr} P P_\phi = \langle\phi|P|\phi\rangle = |\langle\psi|\phi\rangle|^2. \tag{6.89}$$

This is the probability that the projector P has the value 1 in the state $|\phi\rangle$ which is indeed the statistical algorithm of quantum mechanics or Born's rule.

If $\mathcal{P}_{\phi_i}$ are different probability measures on the Hilbert lattice $L = \mathbf{C}(H)$ associated with the normalized vectors $|\phi_i\rangle$ then the convex sum (mixture) $\mathcal{P} = \sum_{i=1}^{n} t_i \mathcal{P}_i$, where $0 \leq t_i \leq 1$ and $\sum_{i=1}^{n} t_i = 1$ is also a probability measure on L. We can immediately compute

$$\mathcal{P}(M_p) = \operatorname{tr} P\rho, \quad \rho = \sum_{i=1}^{n} t_i P_{\phi_i}. \tag{6.90}$$

The convex sum ρ of the one-dimensional projectors P_{ϕ_i} (pure states) is called the density operator (which is generally a mixed state).

Thus, every density operator ρ yields a countably additive probability measure on the subspaces of the Hilbert space. The converse is precisely Gleason's theorem [21], which is a highly non-trivial result in quantum mechanics and its foundations.

Gleason's theorem: Every countably additive probability measure on the subspaces of a separable Hilbert space of dimension greater than two is (i) characterized by a unique non-negative self-adjoint operator ρ satisfying $\operatorname{tr}\rho = 1$ (trace class operator) and (ii) is necessarily of the form

$$\mathcal{P}_\rho(M_p) = \operatorname{tr} P\rho. \tag{6.91}$$

The density operator represents of course the quantum state of the physical system. Gleason's theorem shows therefore that all probability measures on $\mathbf{C}(H)$ are generated by quantum mechanical states.

Gleason's theorem is a substitute to Born's rule. Indeed, if the physical system is prepared in the pure state $|\phi\rangle$ then the density operator which is generally of the form $\rho = \sum_i P_i |\phi_i\rangle\langle\phi_i|$ reduces to the pure state $\rho = |\phi\rangle\langle\phi| = P_\phi$ and we end up with Born's rule (6.89). Thus, we can derive the statistical algorithm of quantum mechanics or Born's rule in a straightforward way from Gleason's theorem.

One immediate corollary of Gleason's theorem is effectively the Kochen–Specker theorem (see the next sections for more detail). Indeed, in Gleason's theorem we are attempting to assign probabilities to closed subspaces of a Hilbert space H such that the probabilities assigned to orthogonal projection operators are additive. This assignment is found to be provided by a density operator ρ. In the Kochen–Specker theorem we will deal instead with the special case in which we attempt to assign only the values 0 and 1 (two-valued probability measures) to the closed subspaces of H in a consistent way (non-contextuality). This assignment, as we will show, is actually impossible. In other words, the set of quantum events $\mathbf{C}(H)$ does not admit a global two-valued probability measure.

This fundamental result can be derived from Gleason's theorem as follows. Every probability measure on $\mathbf{C}(H)$ must be necessarily a continuous mapping into the interval [0, 1]. Indeed, for any density operator ρ the mapping $|\psi\rangle \longrightarrow \mathcal{P}(M_p) = \langle\psi|\rho|\psi\rangle$ (with $P = |\psi\rangle\langle\psi|$) is a continuous function on the unit sphere in H. Obviously, there is no continuous function defined on the unit sphere (because it is a connected space) which takes only the two values 0 and 1. Hence, the set of closed subspaces of the Hilbert space does not admit a two-valued probability measure.

6.4.2 Theorems of quantum/experimental metaphysics

The most important theorems in the foundations of quantum mechanics, which is arguable the central topic in the philosophy of physics, are the Bell and Kochen–Specker theorems. These were the first mathematical theorems in all of metaphysics with Bell's theorem being the most important of the two since it applies to all interpretations of quantum mechanics and only suffers from small loopholes. Quantum metaphysics based on these theorems should then be properly thought of as experimental metaphysics.

Bell's theorem [17], which is called by Stapp [23] the most profound discovery of science, is based on the two assumptions of locality and realism and it is important to all interpretations of quantum mechanics. A theory (hidden variables theory or otherwise) which satisfies these two assumptions will contain certain inequalities which are seen to be violated in quantum mechanics and by Nature (experimentally). Thus, the two conditions of locality and realism are contradictory and typically people tend to abandon the condition of locality, i.e. non-local hidden variables theories of quantum mechanics are ruled out and perhaps Nature itself is non-local.

On the other hand, the Kochen–Specker theorem [18] applies only and specifically to hidden variables theories of quantum mechanics and it relies on three assumptions:

1. *Projection postulate*: This states that physical properties can be mapped one-to-one to projection operators on the Hilbert space. In other words, a yes–no measurement on a physical system can always be devised in such a way as to reveal whether or not the physical system has a certain property. This is a very natural ingredient of the Hilbert space formalism of quantum mechanics.
2. *Value definiteness*: This states that all observables must have definite values at all times. This relies explicitly on the existence of a hidden variables theory of quantum mechanics and it partially reflects the reality condition. Indeed, measurements performed in experiments do not create their outcomes but only reveals pre-existing values.
3. *Non-contextuality*: This states that the outcome obtained in the measurement of an observable does not depend on the order, i.e. the context in which that observable is measured. So a physical property measured in different incompatible measurements should be found to have the same value.

Kochen and Specker showed that the above three conditions are contradictory in any hidden variables theory of quantum mechanics. Typically, people tend to relax the conditions for the Kochen–Specker theory by dropping the requirement of non-contextuality. In other words, hidden variables theories of quantum mechanics and by consequence Nature seem to be contextual, i.e. the context of measurement is crucial.

In this section we will follow the presentation in [24].

Hidden variables theories are meant to provide a completion of quantum mechanics in which the quantum mechanical probabilities, computed from the vector states of the system, are simply epistemic probabilities without any onto-logical import for the state of the system. In other words, hidden variables theories of quantum mechanics are really just Kolmogorov probability theories. The most important assumption underlying hidden variables theories is the second assumption of value definiteness, i.e. that eigenvalues of observables exist independently of any measurement. This assumption is what leads naturally to the third assumption of non-contextuality, i.e. the order of measurement is irrelevant to the value obtained.

von Neumann, in his book [25], was the first to attempt an impossibility theorem against hidden variables theories of quantum mechanics. He considered the case of two self-adjoint operators A and B (not necessarily compatible) which have expectation values $\langle A \rangle = \langle \psi|A|\psi \rangle$ and $\langle B \rangle = \langle \psi|B|\psi \rangle$ in some state $|\psi\rangle$. Each expectation value represents the average obtained in a series of identical measurements performed on the physical system when prepared in the same state $|\psi\rangle$. Let C be the operator given by the linear combination

$$C = \alpha A + \beta B, \quad \alpha, \beta \in \mathbb{C}. \tag{6.92}$$

According to standard quantum mechanics the operator C is also a self-adjoint operator on the Hilbert space with an expectation value in the state $|\psi\rangle$ given by

$$\langle C \rangle = \alpha \langle A \rangle + \beta \langle B \rangle. \tag{6.93}$$

In single measurements of the operators A, B and C we obtain definite values $\nu(A)$, $\nu(B)$ and $\nu(C)$. In a hidden variables theory of quantum mechanics it is assumed that the hidden variables λ (which supplement the state vector $|\psi\rangle$) determine these definite values, i.e.

$$\langle A \rangle_\lambda = \nu(A), \quad \langle B \rangle_\lambda = \nu(B), \quad \langle C \rangle_\lambda = \nu(C). \tag{6.94}$$

The hidden variables λ will be characterized by some probability distribution $\rho(\lambda)$ and the expectation values $\langle A \rangle_\lambda$, $\langle B \rangle_\lambda$, $\langle C \rangle_\lambda$ with respect to this probability distribution are precisely the expectation values $\langle A \rangle$, $\langle B \rangle$ and $\langle C \rangle$, namely

$$\int d\lambda \rho(\lambda) \langle A \rangle_\lambda = \langle A \rangle, \quad \int d\lambda \rho(\lambda) \langle B \rangle_\lambda = \langle B \rangle, \quad \int d\lambda \rho(\lambda) \langle C \rangle_\lambda = \langle C \rangle. \tag{6.95}$$

However, von Neumann assumed the stronger condition that the expectation values $\langle A \rangle_\lambda$, $\langle B \rangle_\lambda$, $\langle C \rangle_\lambda$ satisfy the same condition as the expectation values $\langle A \rangle$, $\langle B \rangle$ and $\langle C \rangle$ and hence the definite values $\nu(A)$, $\nu(B)$ and $\nu(C)$ must satisfy the functional relation

$$\nu(C) = \alpha \nu(A) + \beta \nu(B). \tag{6.96}$$

However, the passage from (6.93) to (6.96) is only valid for compatible observables which are comeasurable. Indeed, for incompatible observables (such as $A = \sigma_x$, $B = \sigma_y$, $C = (\sigma_x + \sigma_y)/\sqrt{2}$) equation (6.96) is simply not true. In fact, hidden variables theories of quantum mechanics are not required to meet the condition (6.96) but only (6.93).

6.4.3 The set-up of the Kochen–Specker theorem

We consider the measurement of the spin components along the Cartesian coordinate axes $(1, 0, 0)$, $(0, 1, 0)$ and $(0, 0, 1)$ of a spin one particle. The measurement in each direction will yield the values -1, 0 and $+1$ with equal probabilities but we cannot measure the spin components in the three directions x, y and z simultaneously, i.e. the corresponding propositions are not comeasurable since the operators J_x and J_y and J_z are not compatible.

However, the squared operators J_x^2, J_y^2 and J_z^2 correspond to compatible observables which can be measured simultaneously. The measurements of the squared spin components along the three perpendicular directions x, y and z correspond therefore to comeasurable propositions and we obtain the two values 0 and 1 with probabilities 1/3 and 2/3, respectively. But the spin one particle must also satisfy the constraint $J_x^2 + j_y^2 + J_z^2 = 2$ and hence the measurement of the squared spin components in any three perpendicular directions will give the answers 1, 0, 1 in some order.

The fact that the three observables J_x^2, J_y^2 and J_z^2 are compatible means that they are commuting as self-adjoint operators on the Hilbert space. This means that there exists a self-adjoint 'Ur'-operator U and three functions f, g and h such that $J_x^2 = f(U)$, $J_y^2 = g(U)$ and $J_z^2 = h(U)$. This is a general result in finite-dimensional Hilbert spaces which states that any set of mutually comeasurable observables (for example the operators in a block) correspond in fact to a single 'Ur'-operator [26]. In the case of the above spin 1 particle the 'Ur'-operator is given explicitly by

$$U = aJ_x^2 + bJ_y^2 + cJ_z^2. \tag{6.97}$$

The operators J_x^2, J_y^2 and J_z^2 can be determined as a function of only the 'Ur'-operator U which defines therefore the functions f, g and h in a closed form. We have explicitly [9]

$$\begin{aligned} J_x^2 &= \frac{1}{(a-b)(c-a)}(U-b-c)(U-2a) \\ J_y^2 &= \frac{1}{(a-b)(b-c)}(U-a-c)(U-2b) \\ J_z^2 &= \frac{1}{(c-a)(b-c)}(U-a-b)(U-2c). \end{aligned} \tag{6.98}$$

The measurement of the 'Ur'-operator U will give the results $a + b$ (associated with $J_x^2 = 1$, $J_y^2 = 1$, $J_z^2 = 0$) or $a + c$ (associated with $J_x^2 = 1$, $J_y^2 = 0$, $J_z^2 = 1$) or $b + c$ (associated with $J_x^2 = 0$, $J_y^2 = 1$, $J_z^2 = 1$). The basic comeasurable propositions of interest are thus given by the following projectors:

- p_1: The measurement result of J_x^2 is 0 (corresponding to a measurement result of U given by $b + c$).
- p_2: The measurement result of J_y^2 is 0 (corresponding to a measurement result of U given by $a + c$).
- p_3: The measurement result of J_z^2 is 0 (corresponding to a measurement result of U given by $a + b$).

Clearly p_1, p_2 and p_3 are compatible projectors on the Hilbert space associated with the measurements of the eigenvalues of the 'Ur'-operator U satisfying thus the completeness relation

$$p_1 + p_2 + p_3 = 1. \tag{6.99}$$

The coordinates axes x, y and z define a tripod and the measurements of the three compatible observables J_x^2, J_y^2 and J_z^2 corresponding to this tripod define an 'element of physical reality' which reduces effectively to the measurement of the eigenvalues of a single 'Ur'-operator U. This measurement yields the values 1, 0 and 1 in some order.

By applying $SO(3)$ rotations we obtain different tripods with coordinates axes $\bar{x}$, $\bar{y}$, $\bar{z}$ associated with different 'Ur'-operators $\bar{U}$. These 'Ur'-operators $\bar{U}$ are incompatible observables. The measurements of the squared spin components along different sets of mutually orthogonal directions given by these rotated tripods define other 'elements of physical reality'. Again, these measurements always yield the values 1, 0, 1 in some order.

The goal then is to assign the numbers $\{1, 0, 1\}$ to *all tripods* (this is the assumption of value definiteness) in the three-dimensional Hilbert space H of the system in *a consistent way* (this is the assumption of non-contextuality which means, in particular, that common axes are assigned the same number). It is sufficient to take H to be $\mathbb{R}^3$ and thus the tripods are nothing but triples of perpendicular axes running through the center of the unit sphere.

We can further convert the problem into a coloring problem by coloring black the axes assigned the number 1 while those assigned the number 0 are colored white. It is expected that 2/3 of the axes should be colored black whereas 1/3 should be colored white. This is bound to fail because if we color, for example, the x-axis white then necessarily the y- and z-axes are colored black but then by applying $SO(3)$ rotations around the x-axis all the rotated $\bar{y}$- and $\bar{z}$-axes are colored black as well. Indeed, a coloring which satisfies these requirements is found to be impossible for finite sets of tripods in $\mathbb{R}^3$ (and does not require the use of the infinite set of all tripods in $\mathbb{R}^3$) which is the content of the Kochen–Specker theorem. Equivalently, this shows that it is impossible to define a two-valued probability measure globally although locally in each block (tripod) a two-valued probability measure exists. Establishing a contradiction in three-dimensional real Hilbert space entails a contradiction in higher dimensional complex Hilbert spaces.

6.4.4 Peres construction

We have now the following concrete coloring problem. We measure the squared spin components J_x^2, J_y^2 and J_z^2 along the three perpendicular directions x, y and z of a spin one particle. The operators J_x^2, J_y^2 and J_z^2 are compatible and therefore we are dealing with comeasurable propositions or observables. The fact that this is a spin one system means that we can only obtain the outcomes 0 (color white) and 1 (color black). The spin one system must also satisfy the constraint $J_x^2 + j_y^2 + J_z^2 = 2$ and hence we are essentially attempting to assign the numbers $\{1, 0, 1\}$ to the tripod (x, y, z). In other words, the comeasurement of the squared spin components along the three perpendicular directions x, y and z of a spin one particle is equivalent to the problem of coloring one leg of the tripod white while the other two orthogonal legs are colored black. This coloring problem is in fact equivalent to the measurement of a single 'Ur'-operator U.

By rotating the tripod (x, y, z) through the application of arbitrary $SO(3)$ rotations we obtain different tripods with different incompatible 'Ur'-operators U.

The goal then is to assign the two colors black and white to all lines through the origin in $\mathbb{R}^3$ in such a way that (i) for any three mutually orthogonal lines one line will be colored white and the other two lines will be colored black and (ii) for any two orthogonal lines one at least is colored black. This coloring is required to be consistent, i.e. every line is colored either black or white and this is the assumption of non-contextuality (the measurement of the squared spin component along any axis does not depend on the order in which the measurements of the tripods to which this axis belong is done). As it turns out, we do no need to consider an infinite number of lines through the origin in $\mathbb{R}^3$ to show the impossibility of such a coloring since a contradiction already arises for a finite number of lines. The most efficient construction of the Kochen–Specker theorem in three dimensions (which is sufficient to establish a contradiction in higher dimensions) is due to Peres and it involves 33 lines [15, 27].

These 33 directions (or 66 if we include the opposite directions) can be given in a coordinate basis as follows [28]. First, we have the nine obvious directions given by:
- The lines through the origin (where the spin one particle lies) which go through the six points $(\pm 1, 0, 0)$, $(0, \pm 1, 0)$, $(0, 0, \pm 1)$.
- The lines through the origin which go through the twelve points $(\pm 1, \pm 1, 0)$, $(\pm 1, 0, \pm 1)$, $(0, \pm 1, \pm 1)$.

These 18 points define nine directions through the origin which intersect the surface of a cube $\mathcal{C}$ of size 2 at the points on the middle of each of the 12 edges plus the points on the center of each of the six faces.

This cube $\mathcal{C}$ has 13 axes of symmetry given by (i) the three directions through the centers of opposite faces, (ii) the four directions along the long diagonals which go through opposite vertices and (iii) the six directions through the midpoints of opposite edges. By rotating $\mathcal{C}$ through 45 degrees about one of its axes (the directions through the centers of opposite faces) we obtain the three cubes $\mathcal{C}_1$, $\mathcal{C}_2$ and $\mathcal{C}_3$ with a total number of axes of symmetry equal to 4×13. Some of these symmetry axes coincide and we are finally left with only 33 independent symmetry axes which are precisely Peres' 33 directions [29]. See also [30, 31] and [32].

An equivalent and much more visual representation of the extra directions of Peres is obtained by considering the largest circle on each face and then drawing the largest squares inside these circles [33]. The extra points of Peres are then the intersection points on these inner squares. See figure 6.4. We obtain then the following extra directions:
- The lines through the origin which go through the 24 points $(\pm 1, \pm \alpha, 0)$, $(\pm \alpha, \pm 1, 0)$, $(0, \pm 1, \pm \alpha)$, $(0, \pm \alpha, \pm 1)$, $(\pm 1, 0, \pm \alpha)$, $(\pm \alpha, 0, \pm 1)$, where $\alpha = 1/\sqrt{2}$. These are the midpoints of the sides of the inner squares.
- The lines through the origin which go through the 24 points $(\pm 1, \pm \alpha, \pm \alpha)$, $(\pm \alpha, \pm 1, \pm \alpha)$, $(\pm \alpha, \pm \alpha, \pm 1)$. These are the vertices of the inner squares.

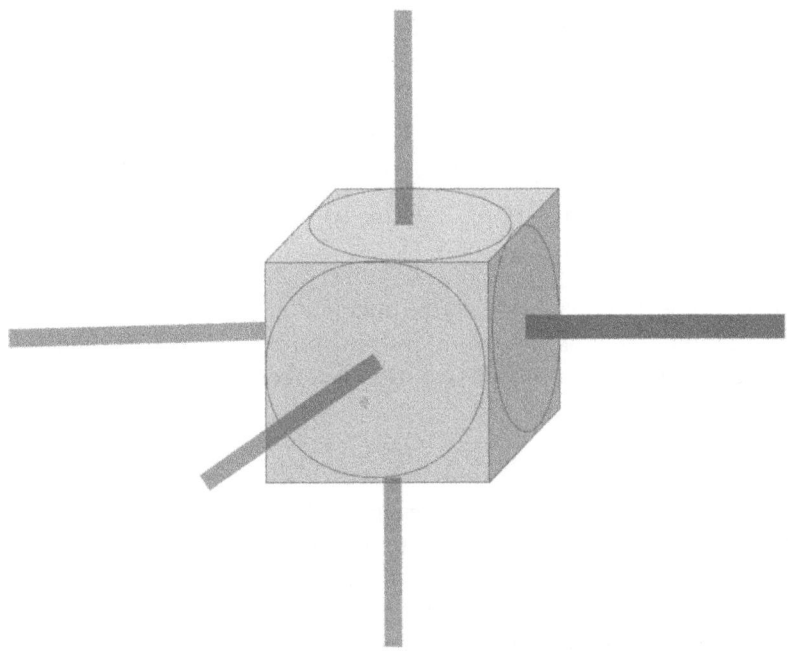

Figure 6.4. The 33 directions of Peres.

Now we attempt to color these 33 lines as discussed above. As it turns this coloring problem admits no solution which is the essence of the Peres proof of the Kochen–Specker theorem. Since this coloring is impossible for these 33 directions it must be impossible for the infinite number of lines going through the origin and since it is impossible in a three-dimensional real Hilbert space it must be impossible in a higher dimensional complex Hilbert space, which is the substance of the Kochen–Specker theorem. The proof consists in producing a contradiction as follows.

1. By rotational symmetry we can start from any point on the cube $\mathcal{C}$. We start with the direction which goes through the point $X = (1, 0, 0)$ and color it with white, i.e. the lines through the points $Y = (0, 1, 0)$ and $Z = (0, 0, 1)$ must be colored black. In fact all orthogonal directions must be colored black. These are given by the 16 points

$$(0, \pm 1, 0), \quad (0, 0, \pm 1), \quad (0, \pm 1, \pm 1), \quad (0, \pm\alpha, \pm 1), \quad (0, \pm 1, \pm\alpha). \tag{6.100}$$

Here we have employed $1 + 2 + 6 = 9$ lines in total.

2. From the six points $(1, \pm\alpha, \pm\alpha)$ and $(0, 1, \pm 1)$ we construct the two tripods

$$\begin{array}{lll}(0, 1, 1), & (1, \alpha, -\alpha), & (1, -\alpha, \alpha) \\ (0, 1, -1), & (1, \alpha, \alpha), & (1, -\alpha, -\alpha).\end{array} \tag{6.101}$$

Obviously, the two points $(0, 1, \pm 1)$ are black. Thus one of the two points $(1, \alpha, -\alpha), (1, -\alpha, \alpha)$ must be black and the other one must be white and also one of the two points $(1, \alpha, \alpha)$, $(1, -\alpha, -\alpha)$ must be black and the other one must be white. The possible pairs of white points are therefore given by the following four possibilities

$$\begin{array}{ll} (1, \alpha, -\alpha), & (1, \alpha, \alpha) \\ (1, \alpha, -\alpha), & (1, -\alpha, -\alpha) \\ (1, -\alpha, \alpha), & (1, \alpha, \alpha) \\ (1, -\alpha, \alpha), & (1, -\alpha, -\alpha). \end{array} \qquad (6.102)$$

By rotating the cube C around the direction going through the point $(1, 0, 0)$ we can easily see that all these four possibilities are actually equivalent. The two resulting white directions can then be taken to be given by the lines going through the two points

$$C = (1, -\alpha, \alpha), \quad C' = (1, \alpha, \alpha). \qquad (6.103)$$

Here we have employed four lines in total.

3. In this last part of the argument we will produce starting from C and C' ten lines leading to a contradiction as follows:
    - The line through the point $D = (\alpha, 1, 0)$ is orthogonal to the line through the point C and since C is white we must color D black.
    - The line through the point $D' = (-\alpha, 1, 0)$ is orthogonal to the line through the point C' and since C' is white we must color D' black.
    - The lines through the points Z, D and $E = (1, -\alpha, 0)$ form a tripod and since Z and D are colored black we must color E white.
    - The lines through the points Z, D' and $E' = (1, \alpha, 0)$ form a tripod and since Z and D' are colored black we must color E' white.
    - The lines through the points E, $F = (\alpha, 1, -\alpha)$, $G = (\alpha, 1, \alpha)$ form a tripod and since E is white we must color both F and G black.
    - The lines through the points E', $F' = (-\alpha, 1, \alpha)$, $G' = (-\alpha, 1, -\alpha)$ form a tripod and since E' is white we must color both F' and G' black.
    - The lines through the points F, F' and $U = (1, 0, 1)$ form a tripod and since F and F' are black we must color U white.
    - The lines through the points G, G' and $V = (1, 0, -1)$ form a tripod and since G and G' are black we must color V white.
    - Finally, the lines through the points U, V are mutually orthogonal but both points U and V are colored white which is *impossible*.

The other ten lines are produced from the other two directions found in step 2 which are given by $(1, \alpha, -\alpha)$ and $(1, -\alpha, -\alpha)$. The Peres configuration is minimal in the sense that the omission of any single line will result in a possible coloring [34].

6.4.5 More on the assumptions of the Kochen–Specker theorem

The main assumptions of the Kochen–Specker theorem are (i) value definiteness and (ii) non-contextuality (the projection postulate is a part of the Hilbert space formalism which cannot be relaxed in any obvious way without affecting the standard framework of quantum mechanics). These two assumptions can be stated slightly differently as follows (see [24] and the extensive references therein).

Let H be a Hilbert space of dimension $D \geqslant 3$ and let $\mathcal{O}$ be a set of observables containing N elements represented by self-adjoint operators on the Hilbert space H. We posit on H and $\mathcal{O}$ the following two very natural assumptions:

1. *Value definiteness*: All elements O_i of $\mathcal{O}$ simultaneously and at all times have definite real values $\nu(O_i)$.
2. *Non-contextuality*: For compatible observables A, B and C we must have the so-called sum and product rules given by

$$C = A + B \Rightarrow \nu(C) = \nu(A) + \nu(B). \qquad (6.104)$$

$$C = A \cdot B \Rightarrow \nu(C) = \nu(A) \cdot \nu(B). \qquad (6.105)$$

These two equations were assumed to hold, in the original consideration of von Neumann, for all observables, which is clearly a much stronger requirement found to be inconsistent with quantum mechanics. As it turns out, even the above weaker requirement of the sum and product rules assumed to hold only for compatible observables is inconsistent with quantum mechanics.

We consider for example an arbitrary self-adjoint operator O in a three-dimensional Hilbert space H which admits three eigenvalues λ_1, λ_2 and λ_3 associated with the three projection operators P_1, P_2 and P_3 on the associated eigenspaces satisfying the completeness relation $P_1 + P_2 + P_3 = 1$.

The projectors P_i are self-adjoint operators on the Hilbert space which are also compatible observables and thus they act as comeasurable 'yes–no' propositions associated with the questions: does the system have the eigenvalue λ_i when we measure O?

By applying the sum rule (non-contextuality) we obtain a constraint on the possible values of these projectors given by

$$\nu(P_1) + \nu(P_2) + \nu(P_3) = 1. \qquad (6.106)$$

The possible values are clearly (i) $\nu(P_i) = 1$ corresponding to the answer 'yes', i.e. the system has the eigenvalue λ_i when we measure O, and (ii) $\nu(P_i) = 0$ corresponding to the answer 'no', i.e. the system does not have the eigenvalue λ_i when we measure O.

Obviously, the choice of the observable O (or equivalently the projection operators P_1, P_2 and P_3) selects a tripod in the three-dimensional Hilbert space and by the assumption of non-contextuality one leg of this tripod must be colored white (answer 'yes') while the other two legs must be colored black (answer 'no').

An arbitrary number n of tripods can be selected by choosing different observables which are generally given by incompatible operators O_1, O_2, ... , O_n on the Hilbert space ($N = 3n$). By the assumption of value definiteness each one of these n tripods must be assigned three values whereas by the assumption of non-contextuality these three values are given exactly by $\{1, 0, 0\}$ (one leg of each tripod is colored white while the other two legs are colored black).

The Kochen–Specker theorem is the statement that the above two assumptions of (i) value definiteness and (ii) non-contextuality are actually contradictory in the sense that the assignment of the values {1, 0, 0} to every tripod in a consistent manner is impossible (a consistent coloring does not exist).

Kochen and Specker in their original proof used a three-dimensional real Hilbert space ($D = 3$) and a set $\mathcal{O}$ of observables with $N = 117$ elements for which the above two assumptions are shown to be contradictory. Peres used also a three-dimensional real Hilbert space ($D = 3$) but a set $\mathcal{O}$ with $N = 33$ elements only and therefore his construction is much more efficient. The fact that the above two assumptions of value definiteness and non-contextuality are contradictory in a three-dimensional real Hilbert space is sufficient to establish that they are contradictory in higher dimensional complex Hilbert spaces.

The sum and product rules are a consequence of the so-called functional composition principle [35, 36], which itself is a consequence of non-contextuality among other assumptions (which are value definiteness, value realism and standard quantum mechanics).

The functional composition principle is the almost obvious statement that every self-adjoint operator must correspond to an observable (in the following we will denote the observable and the corresponding self-adjoint operator by different symbols). More precisely, if **O** is a self-adjoint operator associated with an observable O with value $\nu(O)$ and f is a real function then $\mathbf{f(O)}$ is also a self-adjoint operator associated uniquely with an observable $f(O)$ with value $\nu(f(O))$ given by (where $|\psi\rangle$ is the state of the physical system)

$$\nu(f(O))^{|\psi\rangle} = f(\nu(O))^{|\psi\rangle}. \tag{6.107}$$

By using the statistical algorithm of quantum mechanics (Born's rule) we can instead derive the so-called statistical functional composition principle which states that the above equation holds only statistically, i.e.

$$\text{probability}\,[\,\nu(f(O))^{|\psi\rangle} = a] = \text{probability}\,[f(\nu(O))^{|\psi\rangle} = a]. \tag{6.108}$$

The proof is straightforward (see [24] for example) starting from the Born's rule

$$\text{probability}\,[\nu(O^{|\psi\rangle}) = a_i] = tr P_{|a_i\rangle} P_{|\psi\rangle}. \tag{6.109}$$

The statistical functional composition principle can also be derived from the functional composition principle which makes therefore this latter more plausible. More rigorously, the functional composition principle can be derived from the following principles:

1. *The statistical functional composition principle*: This is given by equation (6.108) and can be derived from the formalism of standard quantum mechanics.
2. *Value definiteness*: All observables simultaneously and at all times have definite real values.
3. *Value realism*: This is a new assumption which is closely related to non-contextuality. It simply states that for every self-adjoint operator there is a well-defined observable. Stated more explicitly, if we can associate an operationally

defined value α (a number built up from a real property) to a self-adjoint operator **O** and if the statistical algorithm of quantum mechanics allows us to define for **O** a probability $\beta =$ probability $[\nu(\mathbf{O}) = \alpha]$ then there must exist an observable O with the value α, i.e. $\alpha = \nu(O)$.

4. *Non-contextuality*: The outcome obtained in the measurement of an observable does not depend on the order in which the measurements of the blocks to which this observable belong is done. Non-contextuality implies that observables O and their values $\nu(O)$ are independent of the measurement context. This independence from context implies a one-to-one correspondence between self-adjoint operators and observables.

We can now derive the functional composition principle from these assumptions as follows. By the assumption of value definiteness an arbitrary physical observable O possesses a value $\nu(O) = a$ from which we can form the value $f(\nu(O)) = b$ for any function f. By using now the statistical functional composition principle we have

$$\text{probability } [f(\nu(O)) = b] = \text{probability } [\nu(f(O)) = b]. \tag{6.110}$$

This means that we have built from the observable O and its value $\nu(O)$ a new self-adjoint operator $\mathbf{f(O)}$ to which we have associated two numbers: a value $\nu(f(O))$ and a probability probability $[\nu(f(O)) = b]$ and hence, by the assumption of value realism, there is a corresponding observable $f(O)$ with value b, i.e.

$$f(\nu(O)) = \nu(f(O)). \tag{6.111}$$

By the assumption of non-contextuality this observable is unique and hence the above last equation is precisely the functional composition principle (6.107).

6.5 CHSH quantum game and free will theorem

We should mention in passing the CHSH quantum game [37] and the free will theorem [30], which can also be viewed as manifestations of Bell's theorem similar to the Kochen–Specker theorem but with different import and implication.

Bell's theorem lies at the heart of quantum philosophy and interpretations of quantum mechanics since it is our single theoretical guiding framework into what we would like to term 'quantum metaphysics'. However, the systematic and detailed explication of Bell's theorem and its quantum metaphysics requires a whole book (if not books) while our discussion of it here in this book does not do it any real justice.

We conclude by noting that quantum metaphysics, as opposed to classical metaphysics, is fully experimental and mathematical with no room for purely rational deliberations if they are not grounded in the scientific method and the formalism.

References

[1] Bell J L and Machover M 1977 *A Course in Mathematical Logic* (Amsterdam: North-Holland)

[2] Hurley P 2000 *A Concise Introduction to Logic* (Belmont, CA: Wadsworth)

[3] Tay S C 2012 *Logic via Algebra* (Lecture notes) https://documents.kenyon.edu/math/SamTay.pdf

[4] Halmos P R and Givant S R 1998 *Logic as Algebra* vol 21 (Washington, DC: Mathematical Association of America)
[5] Fitelson B 2012 *Boolean Algebra* (Lecture notes) http://fitelson.org/
[6] Dalla Chiara M L and Giuntini R 2001 Quantum logic arXiv:0101028 [quant-ph]
[7] de Ronde C, Domenech G and Freytes H 2014 Quantum logic in historical and philosophical perspectives *Internet Encyclopedia of Philosophy*
[8] Birkhoff G and von Neumann J 1936 The logic of quantum mechanics *Ann. Math.* **37** 823–43
[9] Svozil K 1998 *Quantum Logic* (Berlin: Springer)
[10] Wilce A 2017 Quantum logic and probability theory *The Stanford Encyclopedia of Philosophy* **Spring 2017** https://plato.stanford.edu/
[11] Svozil K 1999 Quantum logic: a brief outline arXiv: quant-ph/9902042
[12] Husimi K 1937 Studies on the foundations of quantum mechanics I *Proc. Physico-Math. Soc. Jpn* **9** 766–78
[13] Kalmbach G 1983 *Orthomodular Lattices* (New York: Academic)
[14] Einstein A, Podolsky B and Rosen N 1935 Can quantum mechanical description of physical reality be considered complete? *Phys. Rev.* **47** 777–80
[15] Peres A 1990 Incompatible results of quantum measurements *Phys. Lett.* **151** 107–8
[16] Peres A 1993 *Quantum Theory: Concepts and Methods* (Dordrecht: Kluwer)
[17] Bell J 1964 On the Einstein–Podolsky–Rosen paradox *Physics* **1** 195–200
[18] Kochen S and Specker E 1967 The problem of hidden variables in quantum mechanics *J. Math. Mech.* **17** 59–87
[19] Svozil K and Tkadlec J 1996 Greechie diagrams, nonexistence of measures in quantum logics and Kochen–Specker type constructions *J. Math. Phys.* **37** 5380–401
[20] Kolmogorov A N 1956 *Foundations of the Theory of Probability* (New York: Chelsea)
[21] Gleason A M 1957 Measures on the closed subspaces of a Hilbert space *J. Math. Mech.* **6** 885–93
[22] Greechie J R 1971 Orthomodular lattices admitting no states *J. Comb. Theory* **10** 119–32
[23] Stapp H 1975 Bell's theorem and world process *Nuovo Cimento* **29B** 270–6
[24] Carsten H 2018 The Kochen–Specker theorem *The Stanford Encyclopedia of Philosophy* **Spring 2018** https://plato.stanford.edu/
[25] von Neumann J 1955 *Mathematical Foundations of Quantum Mechanics* (Princeton, NJ: Princeton University Press)
[26] Halmos P 1974 *Finite-dimensional Vector Spaces* (Berlin: Springer)
[27] Peres A 1991 Two simple proofs of the Kochen–Specker theorem *J. Phys. A: Math. Gen.* **24** L175
[28] Fremlin D H 2018 *Peres' 33 Directions* (Lecture notes) https://www1.essex.ac.uk/maths/people/fremlin/n15l31.pdf
[29] Roberts S 2015 *Genius at Play: The Curious Mind of John Horton Conway* (New York: Bloomsbury)
[30] Conway J and Kochen S 2006 The free will theorem *Found. Phys.* **36** 1441
[31] Conway J and Kochen S 2009 The strong free will theorem *Not. AMS* **56** 226–32
[32] Kochen S 2017 Born's rule, EPR, and the free will theorem arXiv:1710.00868 [quant-ph]
[33] Thomas R *Proof of the Kochen–Specker Theorem* (https://plus.maths.org/content/proof-kochen-specker-theorem)
[34] Gould E and Aravind P K 2009 Isomorphism between the Peres and Penrose proofs of the BKS theorem in three dimensions arXiv:0909.4502 [quant-ph]

[35] Redhead M 1987 *Incompleteness, Nonlocality, and Realism. A Prolegomenon to the Philosophy of Quantum Mechanics* (Oxford: Clarendon)
[36] Redhead M 1995 *From Physics to Metaphysics* (Cambridge: Cambridge University Press)
[37] Clauser J F, Horne M A, Shimony A and Holt R A 1969 Proposed experiment to test local hidden variable theories *Phys. Rev. Lett.* **23** 880–4

Chapter 7

Interpretation of the 'Copenhagen interpretation'

In this chapter we take the stronger position that (i) there can be no interpretation of quantum mechanics without the Copenhagen interpretation and (ii) that the Copenhagen interpretation itself requires an interpretation. In particular, it is argued that the complementarity principle is essential to quantum mechanics itself and that as far as the differences between Bohr and Heisenberg go there is a single coherent 'Copenhagen interpretation' which entails both views. In particular the 'classical concepts' for Bohr replace effectively the 'collapse postulate' for Heisenberg. It is the Kantian concept of Bohr's 'classical concepts' which allows us to propose that consciousness is a purely classical concept, creating therefore a dichotomy in Nature between a purely 'quantum Reality' and the purely 'classical cognition' of that Reality. This view is termed 'semi-classical fundamentalism' as opposed to the 'quantum fundamentalism' permeating the current understanding of quantum physics.

7.1 Postulates of quantum mechanics

The philosophy of physics is the philosophy of quantum physics, and the philosophy of quantum physics, also called the interpretation of quantum mechanics or foundation of quantum mechanics, revolves largely around the so-called the 'quantum measurement' problem. Quantum mechanics is the most experimentally successful theory in the history of science (in describing what is observed—the phenomena) but it is also the most speculative theory regarding interpretation and the nature of reality (what is really there—the nomena). Furthermore, quantum mechanics is the most fundamental theory of nature (a stance shared by the majority called 'quantum fundamentalism').

Bohr's interpretation is the original Copenhagen interpretation but it is somewhat different from Heisenberg's standard Copenhagen interpretation outlined, for example, in the two early books by Dirac and von Neumann. Bohr's interpretation is a transcendental, i.e. Kantian, understanding where the so-called 'classical concepts' play the role of Kant's categories of thought (spacetime and causation).

Dyson later proposed understanding the force of gravity as a purely classical phenomenon. Similarly, very early on (almost a century ago), Bohr proposed indirectly and implicitly to understand cognition as a classical phenomenon. Bohr states that for example that the classical concepts define the common language (of classical physics) by which we describe, understand and communicate the results of our experiments and measurements.

Here we would like to propose going one step further and insist that consciousness and cognition are fundamentally classical. Thus, although the world is quantum, the mind is classical and thus the relation between them is necessarily described by the mind in classical terms. In some sense, we are proposing to regard classical mechanics as a fundamental theory of the mind in the same way that quantum mechanics is a fundamental theory of the world.

It is therefore our statement in this chapter that classical mechanics is a fundamental theory on equal footing with quantum mechanics. In other words, there are phenomena and aspects of nature which are genuinely (matter of principle) and/or effectively (matter of practice) classical and not quantum.

The postulates of quantum mechanics are six hypotheses:
- Every state of the physical system is given by a state vector which is a point in a Hilbert space. In the space of configuration the state vector is represented by a wave function.
- Physical quantities are given by Hermitian operators in the Hilbert space.
- The state of the system changes in time in two distinct ways. The first is what von Neumann called 'process II', in which the state of the system evolves unitarily according to the Schrödinger equation. This is a deterministic, continuous and reversible change.
- The second way the state of the system can change in time is what von Neumann called 'process I' and occurs when we perform a measurement on the system. This is an indeterministic, instantaneous and irreversible change. When we measure the physical quantity represented by the operator A we will find as an outcome one of the eigenvalues of this operator with some probability computed from the wave function.

In this measurement (process I) the state of the system changes from the linear superposition of all the eigenstates of the operator A to the eigenstate associated with the eigenvalue obtained in the measurement. We say that the state of the system has collapsed from the linear superposition to this eigenstate we found in the measurement. This empirical result is called the collapse postulate (or reduction postulate or projection postulate).

Thus, before measurement we have a linear superposition (interference between possibilities) whereas after measurement we have a statistical ensemble (without interference). There is no known interaction which accounts for this effect, i.e. how we can go from a pure state to a mixed

state. This is the so-called 'measurement problem' which is very similar to what happens in the information loss paradox in black holes.
- If the physical system S before measurement were in the state $|\psi\rangle$ and the measurement apparatus M were in state $|\phi\rangle$ then the state of the total system $S + M$ will be the tensor product $|\psi\rangle \otimes |\phi\rangle$.
- Let a_i be the eigenvalues of the Hermitian operator A with eigenstates $|a_i\rangle$. If the state of the system before the measurement were given by $|\psi\rangle$ then the probability of obtaining the eigenvalue a_n as an outcome is given by the square of the modulus of the wave function $\langle a_n|\psi\rangle$, i.e. the probability is given by $|\langle a_n|\psi\rangle|^2$. This is Born's statistical rule.

The claim that decoherence solves the measurement problem or the claim that the measurement problem can be subsummed under the heading of the problem of the emergence of classicality (or the boundary between the classical and quantum worlds) are simply an overreach and they are in fact wanting.

Decoherence is not an interpretation but it is a physical process which effectively solves the problem of the emergence of the classical world from the quantum reality.

7.2 From quantum fundamentalism to semi-classical fundamentalism

Quantum mechanics is the most successful theory in the history of science, and it is the most fundamental theory of nature, yet it has what seems to be a fundamental divide between its ability to explain phenomena (what is observed by the senses) and its ability to account for nomena (what is really Reality). This is the problem of interpretation which revolves essentially around the problem of measurement [9]. This discrepancy is perhaps the most severe case of underdetermination in the philosophy of science (a term due to Quine) [10].

Bohr is also the father of the Copenhagen interpretation, a term due to Heisenberg [3] (see also [12]), which is an interpretation far from being a homogeneous view.

Moreover, Bohr is believed to be the father of the so-called 'quantum fundamentalism' [13], which is the dominant view in (theoretical) physics composed of the following two components:
- *The ontological component*: Everything in Nature/the Universe is fundamentally quantum mechanical.
- *The epistemological component*: Everything in Nature/the Universe should be described in quantum mechanical terms.

However, Bohr with his insistence that the wave function does not represent anything real (it is only symbolic) and his widespread use of the notion of 'classical concepts' (which substitute the wave function collapse as we will see) is giving a mostly epistemological interpretation to quantum mechanics, i.e. it goes against the tenets of quantum fundamentalism.

The classical concepts for Bohr are precisely the spacetime and causal descriptions of the physical objects which are needed to characterize reality in an objective way. These concepts are precisely the position, time, momentum and energy which act as the categories of cognition for Kant given, respectively, by space, time, causation and continuity.

The separation between the object and the subject for both Kant and Bohr is a necessary condition for objective knowledge, which is possible for both of them by means of these classical concepts (Bohr) or the spacetime and causal categories of the mind (Kant). In other words, the classical concepts are preconditions of human knowledge.

Thus Bohr's interpretation is neither positivistic nor subjectivistic but it has more affinity with Kant. Other interpretations of Bohr's Copenhagen interpretation (and in particular his use of the classical concepts) run along the lines of empiricism, pragmatism, Darwinism and experimentalism [8].

It is Heisenberg who should really be thought of as the father of quantum fundamentalism since his interpretation of the quantum formalism (the standard Copenhagen interpretation) relies on the two facts that (i) the wave function is a real object (it is pictorial and not symbolic) and that (ii) this wave function goes through collapse in measurement processes (with the collapse being possibly an approximation of a more fundamental unitary process). Thus the description of the measurement apparatus and observers as classical objects by Heisenberg is only a pragmatic and methodological stance.

Our first goal here is to elucidate precisely the differences between Heisenberg's standard Copenhagen interpretation and Bohr's original Copenhagen interpretation. Also, one of our goals here is to take this position (Bohr's more-than-pragmatic insistence on the use of classical concepts and Heisenberg's pragmatic position regarding the measurement process) to its extreme and drop from the theory of quantum mechanics its quantum fundamentalism altogether by insisting that classical mechanics is a fundamental theory for cognition, consciousness, rationality and perception, i.e. for the mental sphere.

Thus although Reality is truly quantum (the ontological component is correct if Reality is identified with Nature/the Universe) our means of accessing/describing it are essentially classical (the epistemological component is certainly not valid). This is what we would like to term 'semi-classical fundamentalism'. In other words, the theory of classical mechanics should be regarded as being as fundamental as the theory of quantum mechanics.

In contrast, quantum fundamentalism implicitly as well as explicitly posits that quantum mechanics is the only fundamental theory of nature and that classical mechanics is only an approximation of quantum mechanics in the limit in which we can neglect Planck's constant. This limit, i.e. $\hbar \longrightarrow 0$ is precisely the classical limit which is also called the correspondence limit. Note that the correspondence principle concerns therefore the theory of quantum mechanics (and not its interpretations) and it is simply the statement that quantum mechanics should reduce in the classical limit to classical mechanics.

The core of the Copenhagen interpretation is the complementarity principle which, for many people, is not a philosophical view but a part of the quantum theory itself. For example, according to Rosenfeld [27] the complementarity principle is the bedrock of the quantum theory and according to Wheeler [7] this principle is the most important scientific revolution of the twentieth century. Thus, for some this principle should not be thought of as only a part of the Copenhagen interpretation but it is in fact an integral part of the quantum formalism itself. This is clearly a very strong statement but it is indeed very difficult to understand quantum mechanics without complementarity.

Although the collapse postulate is the third integral part of the Copenhagen interpretation (in addition to the correspondence and complementarity principles) Bohr strangely never spoke about it in his writings [8]. In other words, there is a Bohr's Copenhagen interpretation which relies only on the complementarity principle and there is a Heisenberg Copenhagen interpretation which emphasizes also the observer and observer-induced wave function collapse.

We claim that these two pictures are equivalent since for Bohr what replaces the collapse postulate is the insistence on the fundamental role of the classical concepts. In other words, Heisenberg operates with the wave function only, which suffers wave function collapse, whereas Bohr operates with the wave function and the classical concepts without collapse. Born's rule is what relates the two entities for both Bohr and Heisenberg.

According to [8] the wave function plays a pictorial role for Heisenberg but it plays only a symbolic role for Bohr. Thus for Bohr the (symbolic) wave function (pure state) allows us to compute the probabilities of outcomes (mixed state) for the classical concepts (physical properties) using Born's rule in the context of a given set of experiments. This seems to be mostly an epistemological rendering of quantum mechanics. Heisenberg insists that the wave function (pure state) suffers collapse toward states distributed according to Born's statistical rule (mixed state) generating as a consequence the observed values with the correct frequencies of the physical observers (physical properties) in the context of a given set of experiments.

The complementarity principle, according to Bohr [2] (see also [8]), plays in the theory of quantum mechanics a similar role to which the principle of relativity plays in the theory of special relativity. For Bohr both quantum mechanics and special relativity showed that observation in physics is contextual, i.e. it requires reference to observers. This is due to a maximum possible velocity (in the theory of special relativity) and a minimum value of the action (in quantum mechanics). These universal limits yield the relativity of simultaneity or time direction (in the theory of special relativity) and the entanglement between objects and observers (in quantum mechanics). In some sense Bohr is insisting that perspectivism (Nietzsche) is not only inevitable but it is in fact natural and fundamental.

The complementarity principle in physics states that complementary observables (which are given by non-commuting operators in the Hilbert space) can only be measured in mutually exclusive experiments and thus the outcomes of these measurements (classical concepts of Bohr) cannot exist simultaneously and moreover they can only exist in reference to some experiment (in stark contrast with classical physics). All these ideas should be (mostly) grounded in Heisenberg's uncertainty principle which Bohr insists rightly on calling it Heisenberg's indeterminacy principle to emphasize the ontological character of this principle as opposed to the merely epistemological content suggested by the word 'uncertainty' [8].

The original complementarity principle (Bohr) is a kinematic–dynamic complementarity principle which states that for stationary states there is a complementarity between the spacetime and causal descriptions, i.e. these two descriptions of the quantum world, although they classically may look contradictory (however, they rest on mutually exclusive experiments) they actually complement each other. For Bohr, the causal (dynamical) description is given by the conservation of energy and momentum (forces) whereas the spacetime (kinematical) description is given in terms of kinematical variables such as position and duration, which are conjugate to the dynamical variables momentum and energy, respectively.

For Heisenberg, there is another type of complementarity between the causal (dynamical) description which is given by the deterministic evolution of the Schrödinger equation (which certainly satisfies the principle of conservation of energy and momentum) and between the spacetime (kinematical) description, which is given in terms of the classical mixed state obtained after the act of measurement.

For Bohr the use of the classical concepts (kinematical or dynamical) is a necessary link between experiments and the formalism of quantum mechanics. However, the quantization of the action due to Planck imposes severe limitations on the use of these classical concepts (which is only allowed in the context of measurement). Indeed, the use of these concepts makes sense only in the context of specific experiments, i.e. particles can be assigned particular values of some (and not all) of the classical concepts only within a specific experiment and thus these values do not correspond to independent properties of the particles.

The physical system is always treated quantum mechanically whereas the measuring apparatus or part thereof is treated classically. However, the quantum mechanical wave function $|\psi\rangle$ provides us, according to Bohr, only with a symbolic (and not a pictorial or literal) representation of reality. Indeed, it allows us to compute the outcomes of measurements of various physical observables, under well-defined experimental conditions, corresponding precisely (in the limit of negligible values of the quantized action) to the classical concepts. Physical observables associated with projection operators which do not commute correspond to complementary classical concepts which can be made manifest in mutually exclusive experiments.

Thus, by means of the classical concepts we can describe all properties of the physical system as they manifest themselves in mutually exclusive measurements which complement each other, thus constructing an objective picture of the physical system as it manifests itself to the senses. These complementary measurements exhibit physical properties of the object which cannot exist simultaneously and can only exist in reference to some experiment (in stark contrast with classical physics).

The mutually exclusive measurements (of incompatible physical quantities) lie at the heart of the principle of complementarity which is, according to Rosenfeld [27], the bedrock of the quantum theory and, according to Wheeler [7], the most important scientific revolution of the twentieth century, and thus this principle should not be thought of as only a part of the Copenhagen interpretation but it is in fact an integral part of the quantum formalism itself.

7.3 The uncertainty/indeterminacy principle

The uncertainty principle was meant by Heisenberg to provide a physical foundation for his matrix quantum mechanics, whereas for Bohr it serves as the mathematical foundation for the complementarity principle. Here, we feel strongly about both perspectives.

The uncertainty principle (or relations) is the statement that conjugate variables (such as position and momentum) cannot be measured exactly and simultaneously, and they can only be determined with uncertainties which cannot be made as small as we desire, but they are (these uncertainties) bounded from below—and by complementarity—by the minimal value of the quantized action, i.e. Planck's constant. In other words, the uncertainty in the position is inversely proportional to the uncertainty in the momentum and hence the exact measurements of the position and the momentum are complementary to each other, i.e. they require mutually exclusive experiments.

In 1925 Heisenberg proposed his matrix quantum mechanics whereas his uncertainty principle was introduced two years later in 1927. In matrix quantum mechanics only observables (physical quantities which are susceptible to measurement) are considered. These observable are infinite dimensional matrices which are identified with Hermitian operators in a separable Hilbert space. For example, the position q and the momentum p are represented by Hermitian operators Q and P in a separable Hilbert space $\mathcal{H}$ which satisfy the canonical commutation relation

$$[Q, P] = QP - PQ = i\hbar, \tag{7.1}$$

where $\hbar = h/2\pi$ is the reduced Planck's constant. The operators Q and P are then, since they do not commute, called incompatible.

The above relation is precisely the quantization of the Poisson bracket $\{q, p\} = 1$ in classical mechanics, i.e. matrix quantum mechanics is formally very close to classical mechanics. Indeed, the correspondence principle states that we must make the replacement

$$q \longrightarrow Q, \quad p \longrightarrow P, \quad \{,\} \longrightarrow \frac{[,]}{i\hbar}. \tag{7.2}$$

The classical limit is then defined by $\hbar \longrightarrow 0$.

However, in 1926 Schrödinger presented his wave quantum mechanics, which is based on state vectors in the Hilbert space $\mathcal{H}$ evolving continuously according to the equation

$$i\hbar \frac{\partial}{\partial t} |\psi(t)\rangle = H|\psi(t)\rangle = \left(-\frac{\hbar^2}{2m}\frac{\partial^2}{\partial q^2} + V(q)\right)|\psi(t)\rangle. \tag{7.3}$$

This equation can be obtained by applying the correspondence principle to the total energy $E = p^2/2m + V$ with the replacement

$$q \longrightarrow Q = q, \quad p \longrightarrow P = \frac{\hbar}{i}\frac{\partial}{\partial q}, \quad E \longrightarrow i\hbar \frac{\partial}{\partial t}. \tag{7.4}$$

In wave quantum mechanics the electron is viewed as a cloud surrounding the nucleus which evolves continuously in spacetime. Thus, the so-called quantum jumps, for example, are not instantaneous as one might think from considerations of matrix quantum mechanics. These quantum jumps are transitions between energy levels (and since these levels are discrete these transitions are truly jumps in the space of energies) and they are clearly not instantaneous in spacetime but are in fact wholly continuous. This picture seems to be fully accepted by Heisenberg who was set therefore to find the analogous physical content (a causal spacetime picture) of his matrix quantum mechanics which culminated in his uncertainty relations.

However, before that Schrödinger himself showed that matrix quantum mechanics (now called the Heisenberg picture) and wave quantum mechanics (now called the Schrödinger picture) are in fact equivalent. The starting point is to remark that in the Heisenberg picture observables $O_H = O(t)$ are dependent on time while state vectors $|\psi_H\rangle = |\psi\rangle$ are not, whereas in the Schrödinger picture observables $O_S = O$ are independent of time while state vectors $|\psi_S\rangle = |\psi(t)\rangle$ are dependent on time. The physically observable expectation values must be the same in both pictures. We write then

$$\langle \psi_H | O_H | \psi_H \rangle = \langle \psi_S | O_S | \psi_S \rangle. \tag{7.5}$$

We take the time derivative of both sides and using the Schrödinger equation we have

$$\left\langle \psi \left| \frac{\partial}{\partial t} O(t) \right| \psi \right\rangle = \frac{\partial}{\partial t} \langle \psi(t) | \cdot O | \psi(t) \rangle + \left\langle \psi(t) \left| O \cdot \frac{\partial}{\partial t} \right| \psi(t) \right\rangle. \tag{7.6}$$

$$i\hbar \left\langle \psi \left| \frac{\partial}{\partial t} O(t) \right| \psi \right\rangle = -\langle \psi(t) | H \cdot O | \psi(t) \rangle + \langle \psi(t) | O \cdot H | \psi(t) \rangle$$
$$= \langle \psi(t) | [O, H] | \psi(t) \rangle \tag{7.7}$$
$$= \langle \psi | [O(t), H] | \psi \rangle.$$

We conclude then that

$$i\hbar \frac{\partial}{\partial t} O(t) = [O(t), H]. \tag{7.8}$$

This is the Heisenberg equation which controls the evolution of time-dependent observables in his matrix quantum mechanics.

In his original derivation Heisenberg considered an operational assumption called the

'measurement = meaning principle'

in [18]. Thus, the meaning of the term 'position of the particle' is tied to the feasibility of a real 'experiment' in which the position of the particle can be measured. If such experiments do not exist then the term 'position of the particle' has no meaning and as a consequence the position of the particle is not a well-defined observable. The same goes for the momentum of the particle. This ('measurement = meaning principle') is then an epistemological principle of a sort.

But does the position or the momentum of the particle exist before measurement? Are these properties real attributes before we measure them?

Heisenberg (according to [18]) seems to employ another principle, the

'measurement = creation principle',

which in his words states that 'I believe that one can formulate the emergence of the classical "path" of a particle succinctly as follows: the "path" comes into being only because we observe it' [6]. This is an ontological principle. Thus, what is real is what is measured.

As a general principle of quantum mechanics (complementary principle) no experiment exists which will allow us a simultaneous precise measurement of two conjugate variables such as position and momentum, which means that the position and the momentum of the particle cannot be simultaneously well-defined. Furthermore, these measurements are actually what created these observables as properties of the particle. This is the origin of the concept of observables in quantum mechanics, which are Hermitian operators in the Hilbert space, which are measurable quantities.

The position is measured, for example, using the so-called Heisenberg's microscope [5] in which light of a very short wavelength (for example γ-rays) shines on an electron of initial momentum $\vec{p}_i$, exhibiting therefore its position with an accuracy $\delta q = \Delta x$ proportional to the wavelength λ. More precisely, we have from the laws of optics the result

$$\Delta x \sim \frac{\lambda}{\sin \theta}. \tag{7.9}$$

The angle θ is the microscope's so-called aperture angle.

On the other hand, the photons of light will collide with the electron in a standard Compton effect in which the electron suffers recoil changing in a unpredictable fashion its momentum to $\vec{p}_f$ within the angle θ. This final momentum cannot be determined in the same experiment (Bohr's complementarity principle). Clearly, the shorter the wavelength λ of the photon the more accurate the measurement of its position will be, but the larger will be its change of momentum $\delta p = |\vec{p}_f - \vec{p}_i|$, which is given (from considerations of the Compton effect) by

$$\delta p \sim \frac{\hbar \sin \theta}{\lambda}. \tag{7.10}$$

From equations (7.9) and (7.10) we obtain the constraint

$$\delta q \delta p \sim \hbar. \tag{7.11}$$

This is Heisenberg's uncertainty principle, which is the physical content of the commutation relation $[Q, P] = i\hbar$. The measurement of the position yields a discontinuous unpredictable change in the momentum (since a measurement is associated for Heisenberg with a discontinuous indeterministic change in the wave function). Thus, an exact precision in the measurement of the position yields an infinite uncertainty (or undeterminacy) in the momentum and vice versa.

Here are the steps conducted in Heisenberg's microscope [18]:

- Heisenberg's 'measurement = meaning principle' allows us to give meaning to the statement 'the initial momentum of the particle is $\vec{p}_i$' (in a previous measurement of the momentum). The momentum is a real property of the particle by Heisenberg's 'measurement = creation principle'.
- At this point the position is not a well-defined observable quantity by the 'measurement = meaning principle' and therefore it is not a real quantity (does not exist) by the 'measurement = creation principle'.
- The statement 'the position of the particle is determined to be at $\vec{x}$ with accuracy $\Delta x = \delta q$' is then given meaning and reality or existence (in the current measurement).
- After this second measurement the momentum of the particle changes as we said in a discontinuous and unpredictable way. According to the 'measurement = meaning principle' this means that the momentum is not a well-defined observable anymore and therefore by the 'measurement = creation principle' it does not exist as real property of the particle.
- This momentum can be measured of course in a third experiment (Bohr's complementarity principle).

The uncertainty principle based on the two principles of 'measurement = meaning' and 'measurement = creation' provide restrictions on possible feasible experiments, on possible objective knowledge, on the actual meaning of language, and first and foremost on the ontological nature of quantum systems.

The uncertainty principle should hold for any pair of conjugate variables which will act as a pair of complementary variables in mutually exclusive experiments. Heisenberg himself considered energy E and time t (but time is not associated with any dynamical variable in quantum mechanics), and action J and angle ω (but action is not associated with any observable quantity in quantum mechanics). In particular, the energy–time uncertainty relation is of fundamental importance to the quantum theory of gravity and it can be given some (practical) sense along the lines of [19] (see also [20, 21]). Strictly speaking one should only consider pair of observable variables which are conjugate (classical mechanics) and complementary (quantum mechanics) to each for the uncertainty principle to hold in a precise sense, for example a pair of two components of the angular momentum in orthogonal directions. The above uncertainty relation for position and momentum were derived from the laws of quantum mechanics in 1927 by Kennard and generalized to any two complementary observables by Robertson in 1929.

For any two complementary observables (given by two Hermitian operators A and B in the Hilbert space $\mathcal{H}$ which are incompatible, i.e. the commutator $[A, B] = A.B - B.A \neq 0$) we can show that

$$\Delta_\psi A \Delta_\psi B \geq \frac{1}{2}|\langle [A, B] \rangle|. \tag{7.12}$$

$\Delta_\psi A$ and $\Delta_\psi B$ are the standard deviations associated with the observables A and B when the system is prepared in the quantum state $|\psi\rangle$. They are defined explicitly by

$$(\Delta_\psi A)^2 = \langle A^2 \rangle_\psi - \langle A \rangle_\psi^2$$
$$= \langle \psi | A^2 | \psi \rangle - \langle \psi | A | \psi \rangle^2 \qquad (7.13)$$
$$= \int dq \psi^*(q) A^2(q) \psi(q) - \left(\int dq \psi^*(q) A \psi(q) \right)^2.$$

The same definition applies for B. The wave function $\psi(q)$ is given by the inner product $\psi(q) = \langle q | \psi \rangle$.

As a general principle of quantum mechanics (complementary principle) no experiment exists which will allow us a simultaneous precise measurement of the two conjugate variables A and B, which means that these two observables cannot be simultaneously well-defined. Furthermore, these measurements are actually what created these observables as properties of the particle.

7.4 The complementarity principle

In the complementary principle the appropriate description of the physical system depends on the experimental context. The physical system interacts with the measurement apparatus and this interaction cannot be determined and is not negligible (in contrast with classical mechanics).

The first hypothesis behind the complementary principle is therefore the so-called quantum postulate [1]. This states that the interaction between the physical system and the measurement apparatus includes at least one Planck's quantum of action. This means that the physical system and the measurement apparatus are entangled and constitute in fact a single system.

However, the results of any experiment must use the concepts and terminology of classical physics, what Bohr calls the 'classical concepts'. This hypothesis is called the buffer postulate in [22] since the 'classical concepts' act as a buffer zone between the quantum description and the experimental realization.

Thus the complementarity principle is based on the two postulates:
- The quantum postulate.
- The buffer (classical concepts) postulate.

Thus if one attempts to measure a given physical observable we can conclude from the first postulate that the interaction between the physical system and the measurement apparatus will include at least one Planck's quantum of action. The physical system is quantum (we cannot take the limit $\hbar \longrightarrow 0$), therefore the interaction between the physical system and the measurement apparatus will include at least one Planck's quantum of action, i.e. it cannot be neglected (in contrast to classical mechanics).

From the second postulate the measurement apparatus can be treated classically (we can take the limit $\hbar \longrightarrow 0$). The physical system despite being quantum must be described in terms of the classical concepts which do not involve the quantum of action, i.e. this interaction cannot be determined (in contrast to classical mechanics). However, this classical description will allow us to talk about the physical system alone. It is in this sense that the measurement defines what can be said about the physical system.

The above description of the measurement of the given physical observable of the physical system (in which the interaction with the measurement apparatus, although it cannot be neglected, is left undetermined) cannot then be extended to the measurements of another physical observable of the physical system which is conjugate to the first one. Indeed, we will face the same problems and these two measurements can only be said to be complementary to each other and the corresponding two complementary descriptions cannot be unified in a single classical description.

The most important example of complementarity in quantum mechanics relates the causal (dynamical) description given by the principles of conservation of energy and momentum and the spacetime (kinematical) description given in terms of kinematical variables such as position and duration, which are conjugate to the dynamical variables' momentum and energy, respectively. These two sets of descriptions are united in classical mechanics but in quantum mechanics they are only complementary.

The other kind of complementarity, which is the kind most often discussed in physics, is the celebrated wave–particle duality (which holds in a single experiment as opposed to the position–momentum complementarity which holds in mutually exclusive experiments). The wave–particle duality states that a quantum system may behave as a particle or as a wave but never as both simultaneously. For example, a wave packet of limited extension in spacetime is in fact a linear superposition of a very large number of plane waves with a large range of frequencies. The uncertainties Δp and ΔE of the momentum and energy of the wave are related to the uncertainties Δx and Δt of the particles by the Heisenberg's uncertainty principle, namely

$$\Delta x \cdot \Delta p \sim h, \quad \Delta t \cdot \Delta E \sim h. \tag{7.14}$$

Thus, the time–energy uncertainty relation can be given a well-defined physical meaning in the case of the wave–particle duality which employs, as Bohr did, Fourier analysis instead of the fundamental commutation relation.

For an extensive recent discussion of the wave–particle duality see [15].

7.5 The correspondence principle

The correspondence principle is a relation between quantum mechanics and classical mechanics which was described by Bohr in three different forms. For the sake of historical authenticity as well as physical accuracy it is better to introduce the correspondence principle within the framework of the old quantum theory (1900–1925) which was developed by Planck, Rutherford, Bohr, Sommerfeld and others as the correct description of the hydrogen atom (in particular) after the sad failure of classical mechanics to account for the behavior of atomic systems.

Here we follow the presentation of [23]:
- First, there was the discovery by Planck in 1900 that radiation from a black body comes in discrete pockets called quanta in contrast to classical electrodynamics where it is assumed that all exchanges of energy are continuous.
- Second, there was the model proposed by Rutherford in 1910 based on his scattering experiments in which the atom is viewed as a large mass centered in the nucleus around which the electrons (with total negligible mass) revolve in circular orbits. This model, as is well known, is inconsistent with classical electrodynamics (the charged electrons will radiate all their energies and fall into the nucleus).
- Bohr in 1913 combined Planck's quantization of the action and Rutherford's planetary model of the atom in three postulates of the old quantum theory:
  - An electron in an atom can only exist in the so-called stationary states which are stable periodic (concentric circular) orbits around the nucleus with no emission of radiation. As long as the electron remains in one of its stationary states it can be described by classical mechanics.
  - The stationary orbits are classically allowed motions which need to satisfy the so-called 'Bohr–Sommerfeld quantum condition' which defines the quantization of the classical action. This is given explicitly by

  $$\oint p_\theta d\theta = nh. \qquad (7.15)$$

  The integral is over a full period of the electron motion, the angle θ is the angle in the plane of the electron's orbit and p_θ is the corresponding conjugate momentum (angular momentum). The integer n is the principal quantum number which characterizes the stationary state.
  - A transition between two stationary states, labeled by the quantum numbers n_1 and n_2, is described by Planck's formula. In other words, these transitions or quantum jumps cannot be described by classical mechanics. Indeed, the frequency of the emitted radiation (photon) is given by the difference in the energy of the two states divided by Planck's constant, namely

  $$\nu_{n_1 \rightarrow n_2} = \frac{E_{n_1} - E_{n_2}}{h}. \qquad (7.16)$$

  This formula is the 'Planck–Einstein–Bohr frequency condition'.
- The old quantum theory of Bohr is therefore the first semi-classical approach in which classical and quantum ideas go hand in hand in solving the problem. It consists in solving the problem using classical mechanics and determining all the possible allowed classical orbits, then imposing the quantum condition to determine the actual allowed quantum motion, and finally identifying radiation with the quantum jumps between the allowed orbits [16].
- The correspondence principle is the requirement that quantum mechanics should reduce to classical mechanics in the classical or correspondence limit $\hbar \longrightarrow 0$. This is the precise definition of the correspondence principle.

However, for Bohr the correspondence principle can mean one of three relations between quantum mechanics and classical mechanics: the frequency relation, the intensity relation and the selection rule relation [23]. As we will see, and to whoever can understand the physics, these three relations are actually equivalent.

- In the frequency relation Bohr states that the single quantum frequency $\nu_{n_1 \to n_2}$ corresponding to the quantum jump $n_1 \to n_2$ between the two stationary states n_1 and n_2 must be equal, for large values of the quantum number n, to the classical frequency $\omega_\tau = \tau\omega$ of the τth harmonic of the classical motion with $\tau = n_1 - n_2$, namely

$$\nu_{n_1 \to n_2} = \omega_\tau = \tau\omega, \quad \tau = n_1 - n_2. \tag{7.17}$$

According to classical mechanics the motion of the electron inside the atom is identified with the motion of an accelerating charged particle in an electromagnetic field. More precisely, in a given orbit the motion is periodic with a fundamental frequency ω and the position of the electron is given by the Fourier expansion

$$\vec{x}(t) = c_1 \cos(\omega t) + c_2 \cos(2\omega t) + c_3 \cos(3\omega t) + \cdots \tag{7.18}$$

The term $c_\tau \cos(\tau\omega t)$ is called the τth harmonic of the classical motion with frequency given by the so-called overtone $\omega_\tau = \tau\omega$. The spectrum of this classical atom is clearly given by the frequencies of the harmonics of the motion, i.e. by $\omega, 2\omega, 3\omega, \ldots$, which are given off in the radiation at once.

In the old quantum theory of Bohr states, on the other hand, it is the quantum jump between the two stationary states n_1 and n_2 that results in a single photon with frequency $\nu_{n_1 \to n_2}$. In an ensemble of atoms undergoing multiple transitions between various stationary states we obtain a spectrum which is evenly spaced only in the limit of large quantum numbers. More precisely, the frequency $\nu_{n_1 \to n_2}$ is statistically equal to the classical overtone $\omega_\tau = \tau\omega$ with $\tau = n_1 - n_2$ in the limit of large quantum numbers. This is the correspondence principle according to the frequency relation.

- The correspondence principle, according to the intensity relation, states that in the limit of large quantum numbers there is again a statistical agreement between the probability of transition or quantum jump between the two stationary states n_1 and n_2 and the square of the amplitude of the τth harmonic of the classical motion with $\tau = n_1 - n_2$, namely

$$P_{n_1 \to n_2} = |c_\tau|^2, \quad \tau = n_1 - n_2. \tag{7.19}$$

This could be viewed as the origin of Born's rule.

- We finish with the correspondence principle according to the selection rule relation. Here Bohr states that a quantum transition or a jump between two stationary states n_1 and n_2 is allowed if and only if there exists a harmonic component of the classical motion with overtone $\omega = \tau\omega$, where $\tau = n_1 - n_2$ and ω is the fundamental frequency. This form of the correspondence principle clearly does not rely on large quantum numbers.

7.6 The philosophy of Bohr

First, with regards to the wave function Bohr believes that quantum mechanics provide only a symbolic view of reality (not literal or pictorial). He can be characterized as an entity realist (who does not commit to anything regarding theory) but not as a theory realist, i.e. for him experiments reveal classical properties of real objects relative to a certain measurement [8].

However, other philosophers characterize Bohr as a theory antirealist, a theory instrumentalist or a phenomenological realist. The first consequence for Bohr is the absence of the collapse of the wave function in his interpretation, which is replaced instead, as we have discussed, by the use of classical concepts.

Regarding the formalism of quantum mechanics itself, i.e. the theory, Bohr is usually interpreted as an instrumentalist (theoretical unobserved entities are logically needed to describe experiments). However, although Bohr insists that the wave function is just a symbol and a tool he does not dismiss its fundamental physical significance altogether. Thus, he shows a number of real and antireal tendencies which go against the early understanding of Bohr along positivistic or subjectivistic lines.

The classical concepts for Bohr constitute the bridge between the symbolism of quantum mechanics and the experimental data obtained in measurements. They are also the means by which we communicate to others what we have done and what we have learned. This is why measurements and observations must be expressed in the language of classical physics.

The modern philosophical understanding of the role of the classical concepts in Bohr's interpretation of the quantum formalism follows five distinct philosophical doctrines:

- *Empiricism*: This is the stance of logical positivism which aims at reducing theoretical entities to terms which only involve sense-data. However, Bohr only aims at describing quantum phenomena using the same concepts (purely theoretical or involving sense-data) as those used in describing classical phenomena. Clearly, the empiricism logic of quantum mechanics itself, of Bohr's original Copenhagen interpretation and of Heisenberg's standard Copenhagen interpretation are quite different from both the logic and empiricism of the logical positivists.
- *Kantianism*: C F von Weizsacker was one of the first thinkers to recognize the similarity between Bohr's complementarity interpretation of quantum mechanics and Kant's transcendental method [25]. Indeed, for Kant objective knowledge requires the so-called transcendental condition, which is constituted by the form of intuition (spacetime) and the *a priori* concepts (called also categories) of understanding and thoughts (such as causation, continuity, unity, plurality and totality). Thus, the mind imposes the categories on sense-data which are then rendered in intuition. These *a priori* concepts of Kant are to be identified with the classical concepts of Bohr such as space, time, momentum and energy. In other words, these classical concepts constitute the transcendental condition for objective knowledge which is restricted in

quantum mechanics by contextuality and perspectivism. The form of intuition (space and time) and the categories of thought (causation, etc) are now conjugate variables which are complementary to each other and can only carry a physical meaning relative to an experiment. In some sense, quantum mechanics according to Bohr's complementarity principle denies any possible objective knowledge about the world in-itself since quantum objects carry physical properties only in the context of actual measurements. This is the 'quantum transcendental method' expressed by the idea of complementarity as stated by von Weizsacker [26]:

'The parallelism of the two attitudes is all the more remarkable, since Bohr never seemed to have read Kant extensively. As distinguished from Kant, Bohr learned from modern atomic physics that there exists science beyond the scope where processes can be described reasonably through properties of objects being independent of the situation of the observer. This is expressed by his idea of complementarity'.

- *Pragmatism*: This is based on the assertion that understanding the physical world at the quantum level requires the objective language provided by the classical concepts which were empirically verified at the classical level. So why change the language if it was efficient before?
- *Darwinism*: Leon Rosenfeld (a close student of Bohr) saw a pivotal role for natural selection in Bohr's thinking [27]. Indeed, it is natural selection that made the human subject an observer who accounts for her/his physical experiences in a common language which is adapted for survival and provided by the concepts of classical physics despite the fact that nature, which is the ultimate arbitrator of reality, is quantum in its core.
- *Experimentalism*: Indeed, Bohr seems to be far more concerned with understanding the quantum results of experiments and observations than with interpreting the formalism of quantum mechanics. But the sharp separation between observer and observed is completely lost in quantum mechanics due to entanglement, which is what Bohr attempts to formalize using the complementarity principle. More precisely, Bohr distinguished between the function (which is always classical) and the structure (which is truly quantum) of the measurement [24]. This functional side of experiments is what necessitates the use of classical concepts to describe them. However, the function of experiments which provide a spacetime description can only be complementary to the function of experiments which provide a causal description. In other words, the classical concepts can be attached to quantum system only in experimental contexts.

The correct philosophical grounding of Bohr's classical concepts seems to be a mixture of Darwinism and Kantianism which may be termed natural Kantianism. This is also discussed in [11]. The classical concepts are essential to the Copenhagen interpretation as stated clearly by Heisenberg [6]:

'The Copenhagen interpretation of quantum theory begins with a paradox. Every physical experiment, no matter if it refers to phenomena of daily life or to atomic physics, has to be described by the concepts of classical physics. The concepts of classical physics represent the language in which we describe the configuration of our experiments and determine the results. We cannot replace them by other concepts. Nevertheless the applicability of these concepts is limited due to the uncertainty relations.'

The importance of the classical concepts lies in the fact that they are indispensable to our relation to nature (which is classical) and not to nature itself (which is purely quantum). Kant has concluded that spacetime and causation are preconditions for perception and experience and thus they are given *a priori* before any interaction with the world. These categories of the mind are precisely the classical concepts.

Natural evolution has shaped the human subject in such a way that its perception requires space and time as preconditions for experience, i.e. survival depends on these classical concepts. These become therefore ontological preconditions not mere epistemological ones as in the transcendental method. See [11] and references therein.

However, quantum mechanics, in contrast to the idealization of classical mechanics, gets closer to the true nature of reality and insists that the use of classical concepts required by the mind and evolution is to be constrained according to the complementarity principle (which is a principle governing the strain between the limitation of the classical concepts and the necessity of using them in describing nature).

Thus, our description of nature (and not nature itself) is determined by the mind in the sense that the categories of the mind (the classical concepts) were evolved to enable survival and not to understand the quantum thing in-itself (which can be understood as either a particle or a wave but its true nature remains inaccessible) and therefore there is a duality between the world and the mind (there is no sharp separation between observer and observed since the quantization of the action cannot be neglected in their interaction), which is encapsulated by the complementarity principle. This understanding of the Copenhagen interpretation is clearly far from being subjectivistic, but it can be re-interpreted along the lines of the perspectivism of Friedrich Wilhelm Nietzsche.

All in all Bohr's theory of complementarity is characterized as an epistemological interpretation of quantum mechanics [8]. It seems to us that Bohr's interpretation is still an ontological interpretation based on semi-classical fundamentalism in the same way that Heisenberg's interpretation is certainly ontological based on quantum fundamentalism.

7.7 An executive summary

We started from the following premises:
- The formalism of quantum mechanics requires an interpretation: there is a discrepancy between what we observe (or the phenomena) and what there is (or the nomena).
- Quantum mechanics, despite being the most successful theory experimentally, suffers from underdetermination.

- An interpretation of quantum mechanics is a philosophical view which attempts to address the measurement problem.
- The Copenhagen interpretation is the original (Bohr), standard (Heisenberg) and historical (Dirac, von Neumann) interpretation.
- The Copenhagen interpretation is the only interpretation which does not require other interpretations for its explication.
- The Copenhagen interpretation requires an interpretation.

The complementarity principle is the bedrock of the Copenhagen interpretation and it is based on the following elements [8, 18]:
- The complementarity principle is due to Planck's quantization of the action which is at the root cause of quantum indeterminism (Heisenberg's uncertainty principle).
- The classical concepts for Bohr constitute the bridge between the symbolism of quantum mechanics and the experimental data obtained in measurements.
- In Bohr's version the wave function is only a symbol and a tool (talking about the collapse of the wave function is therefore meaningless), whereas for Heisenberg (as opposed to being a widespread belief) the wave function is really a real object and its collapse under measurement is a genuine objective process which cannot be analyzed any further.
- Heisenberg also accepts the classical concepts as the required language in which we describe the configuration of our experiments and determine the results. We only need to recall that their applicability is limited due to the uncertainty relations.
- It seems that the wave function for Heisenberg which undergoes collapse under measurement is what replaces the fundamental role of the classical concepts for Bohr, who only thinks of the wave function as a tool and a symbol. The wave function and its collapse are real things for Heisenberg but not for Bohr.
- It is by now clear that philosophers and philosophers of physics (but not as clear for theoretical physicists and physicists) have demarcated the boundary between two quite different versions of the Copenhagen interpretation. The second version, which is the standard Copenhagen interpretation, is due to Heisenberg and it relies (and in fact it has initiated) the so-called quantum fundamentalism, which is a position shared by all other interpretations of quantum mechanics and permeates all of theoretical physics. In contrast, the first version, which is the original Copenhagen interpretation, is due to Bohr and it is really based on what I would like to term semi-classical fundamentalism. Both versions, however, share the complementarity principle, which is truly the bedrock of the Copenhagen interpretation.
- The Copenhagen interpretation is based on the complementarity principle. This principle itself is based on the quantum fundamentalism (for Heisenberg) or the semi-classical fundamentalism (for Bohr).
- Quantum fundamentalism assumes two postulates: (i) the ontological postulate (everything is quantum mechanical) and (ii) the epistemological postulate (everything should be described quantum mechanically).

- Semi-classical fundamentalism assumes that although nature is quantum mechanical, and even perhaps consciousness is quantum mechanical, cognition (the relation between nature and consciousness) is necessarily classical.
- The complementarity principle plays the same role in quantum mechanics that the relativity principle plays in the theory of special relativity. The existence of a minimum value of the action causes observation to be an entanglement problem (and thus contextual and perspectival) in the same way that the existence of a maximum possible velocity causes simultaneity (and thus observation) to be relative.
- The principle of complementarity shows in a spectacular way that in the case where we cannot neglect Planck's quantum of action we cannot have a sharp separation between the quantum physical system under measurement and the measuring instruments (or observer) since they are entangled (form a whole) in contrast once more to the principles of classical physics [17].
- The complementarity principle in quantum mechanics is the natural generalization of causality in classical mechanics. The spacetime and causal descriptions are mutually exclusive.
- The complementarity principle is based on the two following assumptions: (i) the quantum postulate and (ii) the buffer postulate.
- The buffer postulate relies on the vital role of the so-called classical concepts (Bohr) which are replaced by the collapse of the wave function for Heisenberg. Thus Bohr describes the physics by a pair constituted by the wave function and the classical concepts, whereas Heisenberg describes the physics by the wave function alone which suffers collapse under measurement in addition to its unitary evolution.
- These two postulates (quantum and buffer) mean that the interaction between observer and observed is not negligible and cannot be determined. There can be no sharp separation between observer and observed (classical mechanics).
- The classical concepts are the Kantian categories. In fact, Bohr = Kant (epistemology) + natural selection (ontology).
- The mathematical realization of the complementarity principle is given by Heisenberg's uncertainty principle.
- The uncertainty principle is philosophically based on two operational principles: the (measurement = meaning) and (measurement = existence) principles.
- The primary example of the complementarity principle is given by the wave–particle duality.
- Thus, the complementarity principle deals only with complementary variables which can be measured in mutually exclusive experiments. These are precisely the conjugate variables of classical mechanics or the classical concepts of Bohr.
- The classical concepts correspond to complementary observables and thus represents physical properties which are not mutually observed (well-defined, existing) and only exists in the context of mutually exclusive experiments (in contrast to classical mechanics).

- The uncertainty principle is intimately tied to the collapse postulate for Heisenberg. Indeed, the discontinuous and unpredictable changes in the momentum and energy when the position and duration are precisely measured reflects the discontinuous indeterministic change in the wave function under measurement. For Bohr, this discontinuous and unpredictable change in the momentum and energy is not simply a consequence of the discontinuous indeterministic change in the wave function but it also reflects the impossibility of defining precisely the change in momentum and energy when the spacetime position is determined precisely.
- For Bohr the original complementarity is between causation and spacetime, which are precisely Kant's categories of the mind. Hence, for Bohr the semi-classical formulation (Born's statistical interpretation plus classical concepts) is the most fundamental description of natural phenomena, even those involving quantum systems (semi-classical fundamentalism). The complementarity principle is captured mathematically by Heisenberg's uncertainty principle and its most physical realization is the wave–particle duality. But the semi-classical description is captured mathematically by the correspondence principle.
- For Heisenberg the complementarity principle as captured by Heisenberg's uncertainty principle is between conjugate variables with a primary example given by the causation and spacetime variables. The primary physical example of complementarity is again the wave–particle duality. However, Heisenberg also views the deterministic evolution of the wave function and its collapse under measurement as complementarity. For Heisenberg, the collapse of the wave function is a genuine quantum physical process which replaces the role of the classical concepts and hence it is at the origin of quantum fundamentalism. In this case the correspondence principle is a part of the formulation of quantum mechanics and it is not needed in the interpretation.
- The correspondence principle states that quantum mechanics should reduce to classical mechanics in the limit in which Planck's constant is taken to zero.
- Despite the correspondence principle quantum mechanics breaks explicitly and sometimes badly a lot of the assumptions on which classical mechanics is based.
    Some of the assumptions of classical mechanics include:
    – Physical objects, their interactions and the processes of measurements/observations are knowable.
    – A sharp distinction between object (physical system) and subject (observer).
    – Physical objects exist in spacetime regardless of measurement and observation.
    – The intrinsicness of physical properties of a physical object (independent of other objects and measurements).
    – Physical objects separated in spacetime cannot be entangled (cannot be counted as a single object).

- Identical objects can occupy the same spacetime point.
- Determinism.
- Locality.
- Causality.
- Conservation of energy.
- Continuity.

These principles are either modified, broken or badly broken in quantum mechanics. The most spectacular in our opinion is the second principle: there is no sharp distinction between object (physical system) and subject (observer).

- The complementarity principle is also largely misunderstood as a subjectivisitic interpretation. This is far from the truth for both Bohr and Heisenberg. The complementarity principle, viewed through the eyes of epistemology, is saying that quantum mechanics is contextual and perspectivistic, i.e. there is no absolute separation between the observer and observed and there are no measurements from nowhere.
- Bohr is neither a positivistic nor a subjectivistic. He is more likely an entity realist (with regard to the wave function) and instrumentalist (with regard to the formalism) and a natural Kantian (with regard to the more important classical concepts).

References

[1] Bohr N 1928 The quantum postulate and the recent development of atomic theory *Nature* **121** 580–90
[2] Bohr N 1998 Causality and complementarity *The Philosophical Writings of Niels Bohr* ed J Faye and H Folse vol 4 (Woodbridge: Ox Bow) pp 12–29
[3] Heisenberg W 1955 The development of the interpretation of the quantum theory *Niels Bohr and the Development of Physics* ed W Pauli (London: Pergamon) pp 12–29
[4] Heisenberg W 1958 *Physics and Philosophy: The Revolution in Modern Science* (London: Allen and Unwin)
[5] Heisenberg W 1930 *Die Physikalischen Prinzipien der Quantenmechanik* (Leipzig: Hirzel)
Heisenberg W 1930 *The Physical Principles of Quantum Theory* (Chicago: University of Chicago Press) (Engl. transl.)
[6] Heisenberg W 1927 Über den anschaulichen Inhalt der quantentheoretischen Kinematik und Mechanik *Z. Phys.* **43** 172–98
Heisenberg W 1983 *Quantum Theory and Measurement* ed J A Wheeler and W H Zurek (Princeton, NJ: Princeton University Press) pp 62–84 (Engl. transl.)
[7] Wheeler J A 1963 'No fugitive and cloistered virtue'—a tribute to Niels Bohr *Phys. Today* **16** 30
[8] Faye J 2002 Copenhagen interpretation of quantum mechanics *The Stanford Encyclopedia of Philosophy* **Winter 2019** https://plato.stanford.edu/archives/win2019/entries/qm-copenhagen/
[9] Myrvold W 2016 Philosophical issues in quantum theory *The Stanford Encyclopedia of Philosophy* **Fall 2018** https://plato.stanford.edu/archives/fall2018/entries/qt-issues/
[10] Lewis P J 2010 Interpretations of quantum mechanics *Internet Encyclopedia of Philosophy* https://iep.utm.edu/int-qm/
[11] Kober M 2009 Copenhagen interpretation of quantum theory and the measurement problem arXiv: 0905.0408 [physics.hist-ph]

[12] Howard D 2004 Who invented the 'Copenhagen interpretation?' A study in mythology *Phil. Sci.* **71** 669–82
[13] Zinkernagel H 2015 Are we living in a quantum world? Bohr and quantum fundamentalism *One Hundred Years of the Bohr Atom: Proceedings from a ConferenceScientia Danica. Series M: Mathematica et Physica* vol 1 ed F Aaserud and H Kragh (Copenhagen: Royal Danish Academy of Sciences and Letters) pp 419–34
[14] Bohr N and Rosenfeld L 1996 Complementarity: bedrock of the quantal description *Foundations of Quantum Physics II (1933–1958) Niels Bohr Collected Works* ed J Kalckar et al vol 7 (Amsterdam: Elsevier) pp 284–5
[15] Bandyopadhyay S 2000 Welcher Weg experiments and the orthodox Bohr's complementarity principle arXiv:quant-ph/0003073
[16] Jammer M 1966 *The Conceptual Development of Quantum Mechanics* (New York: McGraw-Hill)
[17] Kalckar J, Bohr N, Rosenfeld L, Rüdinger E and Aaserud F 1996 *Foundations of Quantum Physics II (1933–1958)* (Amsterdam: Elsevier) p 210
[18] Hilgevoord J and Uffink J 2001 The uncertainty principle *The Stanford Encyclopedia of Philosophy* **Winter 2016** https://plato.stanford.edu/archives/win2016/entries/qt-uncertainty/
[19] Mandelshtam L I and Tamm I E 1945 The uncertainty relation between energy and time in nonrelativistic quantum mechanics *J. Phys. (USSR)* **9** 249–54
[20] Hilgevoord J 1996 The uncertainty principle for energy and time (PDF) *Am. J. Phys.* **64** 1451H–56
Hilgevoord J 1998 The uncertainty principle for energy and time. II *Am. J. Phys.* **66** 396–402
[21] Busch P 1990 On the energy–time uncertainty relation. Part I: Dynamical time and time indeterminacy *Found. Phys.* **20** 1–32
Busch P 1990 On the energy–time uncertainty relation. Part II: Pragmatic time versus energy indeterminacy *Found. Phys.* **20** 33–43
[22] Scheibe E 1973 *The Logical Analysis of Quantum Mechanics* (Oxford: Pergamon)
[23] Bokulich A and Bokulich P 2010 Bohr's correspondence principle *The Stanford Encyclopedia of Philosophy* **Fall 2020** https://plato.stanford.edu/archives/fall2020/entries/bohr-correspondence/
[24] Camilleri K 2017 Why do we find Bohr obscure? Reading Bohr as a philosopher of experiment *Niels Bohr and the Philosophy of Physics* ed J Faye and H Folse (London: Academic) pp 19–48
[25] von Weizsäcker C F 1966 Kant's theory of natural science according to P Plaass *Kant's Theory of Natural Science* ed P Plaass (Dordrecht: Kluwer) pp 167–87
[26] von Weizsaecker C F 1971 *Die Einheit der Natur (The Unity of Nature)* (München: Carl Hanser)
[27] Rosenfeld L 1961 Foundations of quantum theory and complementarity *Selected Papers of Léon Rosenfeld* ed R S Cohen and J J Stachel (Dordrecht: Reidel) pp 503–16

Chapter 8

Neutral monism, perspectivism and quantum dualism

In this chapter we revisit quantum mechanics in the Wigner–von Neumann interpretation, which is the most vocal among all the Copenhagen variants with regard to observer-participancy and/or observer-determinacy. This interpretation, as we will see, is characterized by (i) a quantum dualism between matter and consciousness unified within an informational neutral monism, (ii) a quantum perspectivism which is extended to a complementarity between the Copenhagen interpretation and the many-worlds formalism, (iii) a psychophysical causal closure akin to Leibniz parallelism and (iv) a quantum solipsism, i.e. a reality in which classical states are only potentially existing until a conscious observation is made.

In our view the most important characteristic of the complementarity principle, of the Copenhagen interpretations (such as the Wigner–von Neumann) and of quantum mechanics itself is 'quantum perspectivism', which bears an intriguingly strong resemblance to Nietzsche's perspectivism (or more precisely Putnam's internal realism). In this chapter we will also discuss, among many other things, two fundamental topics of interest to the philosophy of quantum physics, which are (i) the Boltzmann–Schuetz hypothesis and (ii) the mind/body problem.

8.1 The anthropic principle

In 1938 Dirac [1] noted that the ratio between cosmological constants and atomic constants gives large numbers of equal magnitude. For example, the ratio of the electromagnetic force to the gravitational force between an electron and a proton is

$$N_1 = \frac{F_C}{F_G} = \frac{e^2}{4\pi\epsilon_0 G m_p m_e} \sim 10^{39}. \tag{8.1}$$

Similarly, the ratio of the radius of the Universe $R_U = c.t_U$ to the radius R_e of the classical electron is also of this order, namely

$$N_2 = \frac{R_U}{R_e} = \frac{c \cdot t_U}{R_e} \sim 10^{40}, \quad R_e = \frac{e^2}{4\pi\epsilon_0 m_e c^2}. \tag{8.2}$$

Also, the ratio t_U/t_e of the age of the Universe to the time $t_e = e^2/m_e c^3$ required for light to cross an atom is of the same order, and the square root of the ratio of the mass $M_U = c^3 t_U/G$ of the Universe to the mass m_p of the proton is of the same order of 10^{40}.

These coincidences, such as $N_1 = N_2 \sim 10^{40}$, are interpreted by Dirac objectively, i.e. no properties of observers should be included in the explanation. Thus these numerical coincidences are not approximate but are exact equations related to the age of the Universe which should hold at all times, i.e. the dimensionless number 10^{40} is large because the age of the Universe is large.

This interpretation lead Dirac to the conclusion that the gravitational constant G should depend on time (the age of the Universe) t_U as

$$G \sim 1/t_U. \tag{8.3}$$

This was not confirmed experimentally and in fact it contradicts the principle of conservation of energy in general relativity.

However, in 1961 Dick [2] interpreted these coincidences only anthropically, i.e. the observers play a privileged role (very reminiscent of quantum mechanics) which turns out to be the correct approach. Thus these numerical coincidences are objectively observed (not objectively real) only because the very narrow range of values of the physical constants, which lead to these observed coincidences, is the only range consistent with the evolution of life and the existence of intelligent life capable of observing the Universe. In other words, these coincidences are only approximate relations related to the fact that intelligent life (observers) can only exist at this epoch of the history of the Universe. This is the first successful application of the anthropic principle.

The two competing explanatory projects are therefore acausal anthropic (the observer is essential) and causal objective (no role for the observer). We have then:

- *Dirac*: These coincidences, such as $N_1 = N_2 \sim 10^{40}$, are interpreted by Dirac objectively. This leads to the conclusion that the gravitational constant G should depend on time (the age of the Universe) t_U as

$$G \sim 1/t_U. \tag{8.4}$$

- *Dick*: These coincidences are interpreted anthropically, i.e. the observers play a privileged role (very reminiscent of quantum mechanics).

  Thus these numerical coincidences are objectively observed (not objectively real). The observed narrow range of values of the physical constants is the only range consistent with the evolution of life and the existence of intelligent life capable of observing the Universe.

Dicke in 1957 [3] and then in 1961 [2] gave a very compelling interpretation, based implicitly on the anthropic principle, of the observed numerical coincidences obtained when taking the ratio of cosmological constants to atomic constants.

He reasoned that the age of the Universe at the current epoch cannot be random but it is exactly such that life and intelligent life can evolve 'now' at the current epoch so that it can observe this Universe. In other words, any change in the physical constants of this Universe will prevent the existence of rational creatures who can observe it.

The Universe is in fact in its golden age. If the Universe were ten times younger then carbon levels would not have sufficient time to build up by means of nucleosynthesis, whereas if the Universe were ten times older most stars would have left the main sequence, i.e. became too old to sustain life. Thus, the existence of life and intelligent life can only occur during the golden age of the Universe, i.e. 'now' or the current epoch. The anthropic principle (in its strong form since the weak form is rather trivial) is seen as a relaxation of the Copernican (cosmological) principle which states that human observers are not central in the Universe and they do not occupy a special or privileged location in the Universe.

In 1973 Brandon Carter put forward the anthropic principle in its explicit modern form by stating that 'although our situation is not necessarily central it is inevitably privileged to some extent' [4]. Thus, although human observers are not central in the Universe they are certainly special or privileged in the sense that the Universe is uniquely fine-tuned for their existence only at the present epoch. A fine-tuned universe means a universe with a set of physical constants occupying a very narrow range of values consistent with the existence of life and intelligent life.

This coincidence between the perfectly fine-tuned values of the physical constants and the emergence of intelligent life is then not random but it is in fact necessary for the emergence of this intelligent life who can observe these fine-tuned values. For example, a slight increase in the strength of the strong interaction would have converted all hydrogen in the early Universe into helium and as a consequence water and stars which are necessary for life would not have emerged [5].

The anthropic principle is then really a hypothesis about the non-constancy of the physical constants which only makes sense if there is a 'universe ensemble' which is nowadays exemplified by the concept of the 'multiverse' of eternal inflation (which is also intimately related to the concept of the string landscape of vacuum states). The physical constants, and perhaps even the physical laws, may change across the universes and the strong form of the anthropic principle becomes effectively a selection effect among the universes of the multiverse (in the same way the weak anthropic principle is a selection bias among spacetime regions of the same universe). Only the universes that have values of physical constants that allow the emergence of observers are observed. Thus, the anthropic principle shows that observations by intelligent observers necessarily involve the observer selection bias or effect, which is the fact that the observed universe must be consistent with the existence of intelligent observers who perform these observations.

Is this argument a circular argument? Answer: No.
- A fine-tuned universe means a universe with a set of physical constants occupying a very narrow range of values consistent with the existence of life and intelligent life.
- The physical constants are not constants—a 'universe ensemble' given by the multiverse of eternal inflation.

- Thus, the anthropic principle shows that observations by intelligent observers necessarily involve the observer selection bias or effect which is the fact that the observed universe must be consistent with the existence of intelligent observers who perform these observations.

Hence the anthropic principle acquires a real force by adjoining to it the assumption of the multiverse of eternal inflation (see for example [6] for a simplified account).

The inflationary era starts moments after the Big Bang. In the case of normal expansion at less than the speed of light the whole universe remains within our horizon. However, during inflation the Universe undergoes a rapid exponential expansion faster than the speed of light (spacetime itself expands not matter which makes this consistent with special relativity). Inflation is a symmetry breaking phase transition occurring at the scale of the grand unified theory (GUT scale), where an enormous amount of energy is released producing an overpressure causing space-time to expand exponentially by a factor of 10^{54}.

As a consequence the observable universe which is the part within our horizon is only a bubble inside the whole universe. There is, however, an infinite number of other bubbles lying outside the horizon of our observable universe and hence they are causally disconnected from our bubble for all times in the future. The ensemble of all these bubble universes is the multiverse and the number of bubbles in eternal inflation is infinitely growing (inflation is a continuous process in the multiverse). Each bubble is a different universe with its own different physical constants (and its own different physical laws if we also adjoin to eternal inflation the string landscape with its 10^{500} vacuum states each with its own distinct low-energy physics).

According to the anthropic principle, some of these bubble universes will not evolve life, some of them will evolve life, and some of them will evolve intelligent life like the one we find in our own bubble. The many-worlds formalism of quantum mechanics adds the property of linear superposition between the different quantum states of the same bubble universe and even between the different bubble universes of the multiverse.

8.2 Quantum dualism as an informational neutral monism

A revamping of the Wigner–von Neumann interpretation [7, 8] is presented here which involves in an essential way (i) perspectivism [9] and quantum logic [10], and (ii) naturalistic dualism [11] (quantum matter and classical consciousness are distinct aspects of a more neutral substance, i.e. neutral monism). The resulting quantum dualism involves an extension of Bohr's complementarity [12] to a principle of complementarity between the local first-person subjective observers of the Copenhagen interpretation and the global third-person objective observers of the many-worlds formalism.

The collapse of the wave function (which is tied to the arrow of time and a compatibilistic account of free will) is seen as a psychophysical force connecting the physical and mental, similar (but acting in the opposite direction) to the psychophysical force producing qualia. The relation between the physical and the mental is modeled on the many-minds interpretation [13], producing therefore a mental causation similar to Leibniz parallelism (see [14] and references therein). This form of causation produces also a weaker form of solipsism consistent with Bell's theorem [15, 16]. Indeed, it is argued that this quantum dualism is characterized by an apparent subjective idealism (solipsism) in which the consciousness of the first-person observer causes the potentially existing classical (pointer) states to actualize only in the sense of psychophysical parallelism.

In summary, this effective quantum dualism (with its fundamental description as an informational neutral monism) is quite different from both the classical Cartesian dualism [17] and the classical Spinozian neutral monism [18]. It seems to avoid the two main problems of classical physicalism [19]:
  i. the problem of mental causation (the many-minds formulation and psychophysical parallelism) and
  ii. the hard problem of consciousness (naturalistic dualism).

However, this single-world quantum dualism is dual (in virtue of quantum perspectivism) to a pure physicalism in a many-worlds formalism where the Heisenberg cut [20] is placed at infinity, unitarity is the only law and there is no collapse.

8.3 The collapse of the wave function and the measurement problem

The Schrödinger's cat experiment [21] is perhaps the most illuminating thought experiment which can be used to delineate precisely what the measurement problem in quantum mechanics really is. The physical system under consideration consists of a radioactive atom plus a cat enclosed within the wall of a room together with a first-person observer called Schrödinger standing outside the room.

A third-person observer (let us call him Wigner) is standing outside a larger room which encapsulates the Schrödinger's cat experiment, i.e. we are in fact considering a variant of the Wigner's friend experiment [22] where Schrödinger plays effectively the role of the friend.

A typical quantum process consists then of the following four stages:
1. The initial state of the observer outside the room is $|happy\rangle$, of the cat inside the room is $|alive\rangle$ and if the radioactive atom is $|undecayed\rangle$.
2. After one hour there is a fifty percent chance that the radioactive atom will decay thus activating a mechanism which releases a poison that kills the cat instantly. The states of the system radioactive atom + cat after one hour is given therefore by the entangled state

$$|\psi\rangle = \frac{1}{\sqrt{2}}(|0\rangle + |1\rangle), \qquad (8.5)$$

where

$$|0\rangle = |alive\rangle|undecayed\rangle, \quad |1\rangle = |dead\rangle|decayed\rangle. \qquad (8.6)$$

The entangled state $|\psi\rangle$ means that a microscopic event (the decaying of the radioactive atom) is amplified to a macroscopic event (the life or death of the cat) thus the linear superposition principle which is known to be experimentally satisfied in all quantum situations is transferred to classical scales where only one branch of the wave function is observed to be realized at any one time. Indeed, the cat can be thought of as performing a measurement on the atom and finding it decayed, and hence the cat dies in one branch of the wave function, or finding it undecayed, and hence the cat survives in the other branch of the wave function.

3. In the third stage the observer (Schrödinger) makes his measurement on the state of the entangled system atom + cat by entering the room and looking at the cat thus becoming himself entangled with it. The complete entangled state is again given by a linear superposition of the form (8.5), namely

$$|\Psi\rangle = \frac{1}{\sqrt{2}}(|0\rangle + |1\rangle). \tag{8.7}$$

But the two branches $|0\rangle$ and $|1\rangle$ are now given by the states

$$|0\rangle = |happy\rangle|alive\rangle|undecayed\rangle, \quad |1\rangle = |sad\rangle|dead\rangle|decayed\rangle. \tag{8.8}$$

Thus, in the branch $|0\rangle$ the observer is happy to see that the cat is alive because the atom did not decay whereas in the other branch $|1\rangle$ the observer is sad to see the cat is dead because the atom decayed. These two states are maximally entangled and that is why the state $|\Psi\rangle$ is called a pre-measurement state. Indeed, this pre-measurement state contains coherence and interference between the branches and the pointer states $|0\rangle$ and $|1\rangle$ (which form the preferred-basis observed at the classical and macroscopic levels) are still not actually realized. In fact, these states effectively do not exist before the completed measurement [15, 16]. The coherence and interference between the branches can be seen explicitly from the pure density matrix ρ associated with the pure state $|\Psi\rangle$ given explicitly by

$$\rho = |\Psi\rangle\langle\Psi| = \frac{1}{2}|0\rangle\langle 0| + \frac{1}{2}|1\rangle\langle 1| + \frac{1}{2}|0\rangle\langle 1| + \frac{1}{2}|1\rangle\langle 0|. \tag{8.9}$$

The last two terms are the interference terms and interference, as poignantly stated by Feynman, is the hallmark or even the mother of all quantum behavior.

From the perspective of the third-person observer (Wigner) the conscious states (happy, sad) of the first-person observer (Schrödinger) do not effectively exist before reducing the wave function (8.7) in the same way that the states (alive, dead) of the classical cat do not exist until the measurement is completed. Thus, the first-person observer in the pre-measurement state acts in some sense as a philosophical zombie. In the words of Wigner his friend Schrödinger is 'in a state of suspended animation' before the measurement, i.e. while the superposed state $|\Psi\rangle$ given by (8.7) remains coherent.

4. The fourth and last stage is the completed measurement which is described by the reduced density matrix

$$\rho_r = \frac{1}{2}|0\rangle\langle 0| + \frac{1}{2}|1\rangle\langle 1|. \tag{8.10}$$

Thus, the off-diagonal elements responsible for interference are canceled and we end up with ordinary probabilities for mutually exclusive events, i.e. either the cat is alive or is dead with probability equal one half for each. However, this is a mixed density matrix since there is no state $|\Psi\rangle$ in the Hilbert space for which $\rho_r = |\Psi\rangle\langle\Psi|$.

The completed measurement is therefore given by the (discontinuous, irreversible, instantaneous, non-deterministic and non-unitary) transition

$$\rho \longrightarrow \rho_r. \tag{8.11}$$

This is the collapse or reduction postulate and there is no known process in nature which effectuates this transition explicitly. This is the measurement problem. Decoherence [23] tries to effectuate a collapse-like transition by means of a unitary process which couples the system to the environment, but this is beside the point since our fundamental working assumption here is that the above process (8.11) is necessarily non-unitary with respect to the first-person observer.

We have thus two fundamental quantum processes, as formulated originally by von Neumann in his book [7]:
- Process I which is given by the collapse of the wave function (occurring in stage four).
- Process II which is given by the unitary evolution in time generated by a Hamiltonian H, i.e. it is given by the Schrodinger equation (controlling stages one, two and three).

The Copenhagen interpretation is a broad interpretative framework of quantum mechanics due originally to Bohr [12], which can be characterized mainly by the assumption that the collapse of the wave function is a genuine independent and fundamental process in nature (a fifth force of a sort) which is not reducible to any of the other known interactions. This assumption is of course also shared by many other interpretations of quantum mechanics.

The weaker assumption that the collapse postulate is an approximation to some more fundamental unitary law of nature is the main characterization of the many-worlds formalism [24] and similar interpretations. See also [25].

These two standpoints are generally taken to be mutually exclusive interpretations of the laws of quantum mechanics, which is quite an unfortunate situation. Indeed, it is always assumed either explicitly or implicitly that the collapse of the wave function and the subsequent actualization of the classical states of the cat as seen by the first-person observer (Schrödinger) is a fact in gross contradiction with the branching of the wave function, i.e. the real existence of a coherent linear superposition of the classical states of the cat as seen by the third-person observer (Wigner).

The central thesis of this chapter is precisely the converse claim, i.e. that the Copenhagen interpretation and the many-worlds formalism are not contradictory but they are in fact complementary. The Copenhagen interpretation provides the local perspective of reality whereas the many-worlds provides the global perspective or view of reality. This is similar to the black hole complementarity principle in which the descriptions given by the asymptotic (Schwarzschild) and infalling observers, although they may look contradictory classically, are in fact complementary semi-classically.

Thus, a more complete and comprehensive interpretation of quantum mechanics admits the collapse of the wave function as a fundamental process of nature brought about by the observation of first-person local (or simulated as we will discuss in the conclusion) observers but also admits that the laws of nature are fully unitary, i.e. the branching of the wave function and the superposed states are real effects in the world with respect to third-person global (unsimulated) observers. It only seems that the world is full of the first kind of observers who exist in a local system of coordinates not allowing them to access easily the full structure of the 'curved manifold' of reality. See also Tegmark for example in [26].

An independent argument for the complementarity relation between the Copenhagen interpretation and the many-worlds formalism is given by the thought experiment considered by Susskind in [27] which involves, among other things, the properties of entanglement entropy and black holes.

8.4 Quantum perspectivism and quantum logic

The observer plays a fundamental role in quantum mechanics. The first-person observers of the Copenhagen interpretations (which dominate the world) provide the local view of reality while the third-person observers of the many-world formalism provide the global view of reality.

Therefore, quantum mechanics is strongly perspectival in character, which was formalized using the language of quantum logic in [9]. In philosophy, perspectivism is the view due originally to Nietzsche in which it is maintained that all reality is actually perspectival, i.e. there is no objective reality out there independent from the knowing subject and free from interpretations and perspectives [28] (see also Leibniz and his theory of monads [14]). Perspectives for Nietzsche provide an 'optics' of knowledge and they constitute the fundamental condition of the conscious observer in his search for value and meaning in existence and life. Somewhat more precisely, perspectivism can be interpreted as a middle position between metaphysical realism and relativism akin to Putnam's internal realism [29].

For our purposes, every physical theory is characterized by a certain logic $\mathcal{L}$. For example, the logic of classical mechanics is a Boolean algebra $\mathcal{L}$ which is an orthocomplemented distributive lattice based on the power set of the phase space Σ.

In other words, classical events (also called experimental propositions) are subsets of the phase space, pure states semantically decide the truth value of experimental propositions, and the corresponding Boolean algebra underlies Kolmogorovian probability theory.

On the other hand, it was shown by Birkhoff and von Neumann in their seminal paper [10] that the logic $\mathcal{L}$ underlying quantum mechanics is given by the Hilbert lattice of projection operators on the Hilbert space **H**, which is a non-Boolean, non-distributive and orthocomplemented lattice. Therefore the quantum events (or experimental propositions) are given in this case by closed linear subspaces of the Hilbert space, i.e. by projection operators, while pure states decides the truth value only probabilistically, and the corresponding Hilbert lattice or logic of projectors underlies the standard Born's rule.

The differences between the Boolean algebra of the phase space and the Hilbert lattice of projectors stems mostly from the logical disjunction operation **OR**, which in the classical case is given by the union of subsets of the phase space, whereas in the quantum case it is given by the direct sum of closed linear subspaces of the Hilbert space.

The problem of the classical limit in this context is seen as the problem of which projectors and their corresponding closed linear subspaces will tend in the limit $\hbar \longrightarrow 0$ to localized subsets of the phase space. This is clearly a highly non-trivial problem since the vast majority of projectors in the Hilbert space will certainly fail to admit any recognizable classical limit.

In the classical case there is therefore a single perspective (corresponding to the Boolean structure of the classical logic) relative to which we can observe every possible measurable property of the system but in the quantum case there is no single privileged perspective but instead there is an intricate web of non-trivially interlocked classical perspectives, each of which corresponds to a maximal Boolean subalgebra (also called a block) of the Hilbert lattice. A block corresponds naturally to a maximal number of compatible and therefore comeasurable observables which can only be defined locally but not globally. In other words, every block is only locally measurable, i.e. it cannot be defined globally independently of the simultaneous measurement of the other blocks.

The Hilbert lattice in two dimensions can be viewed as a non-trivial disjoint union (pasting) of blocks. However, by Gleason's theorem [30] and its generalization the Kochen–Specker theorem [31] the Hilbert lattice does not admit in general a Boolean reduction, i.e. there is no homomorphism from the Hilbert lattice into a Boolean algebra or to a disjoint collection (pasting) of Boolean algebras which is perhaps the best characterization of the measurement problem. It is worth pointing out that the insistence on the Boolean character is precisely the assumption of reality envisaged by the EPR argument [32].

Thus, in quantum mechanics the Hilbert lattice admits a decomposition into maximal Boolean subalgebras or blocks which define an intricate web of non-trivially interlocked classical perspectives. These perspectives are complementary to each other and their totality defines an omni-perspective (which is seeing everything from everywhere as defined by Nietzsche originally) which is the maximal possible perspective allowed by quantum mechanics. These perspectives are associated naturally with the first-person observers of the Copenhagen interpretation.

The perspective of the third-person observer of the many-worlds formalism which sees coherent linear superpositions of classical states (such as dead and alive cats) is a non-perspective (which is seeing everything from nowhere, i.e. a God's eye view of a sort, which is again Nietzsche's terminology) which is logically impossible according to Nietzsche. However, this impossibility is simply due, as we have seen, to the fact that the world is full of first-order observers who are not directly aware of coherent linear superpositions, i.e. consciousness is through and through classical, which is our second most important thesis in this essay.

Hence the complementarity between the many-worlds formalism and the Copenhagen interpretation proposed in this essay is nothing other than an extension of Bohr's complementarity principle which holds among first-person observers providing classical perspectives on the world [12].

8.5 Physicalism, naturalistic dualism and neutral monism

The Heisenberg cut [20] which demarcates the boundary between the observer (classical) and the observed (quantum) plays for the first-person Copenhagen observer a role similar to the role played for the asymptotic Schwarzschild observer by the event horizon (which separates the inside and outside of the black hole). Indeed, the placement of the cut, although arbitrary, should be thought of as separating the accessible quantum degrees of freedom (associated with the observed physical system or with the entirety of the physical universe) from the inaccessible classical degrees of freedom (those associated with the observer or more precisely with her consciousness) in the same way that the event horizon separates the accessible degrees of freedom (those outside the horizon) from the inaccessible degrees of freedom (those inside the horizon). Thus, the fundamental quantum duality proposed in this essay between observer and observed is really a duality between a strictly quantum physical universe and an exactly classical consciousness. This is a naturalistic dualism not necessarily a Cartesian one, as we will now elucidate further.

The location of the Heisenberg cut is quite arbitrary under the hypothesis of physicalism which entails in particular the assumptions (i) of material monism and (ii) that the physical world is causally closed. Indeed, if there is nothing else but the material substance then there is no intrinsic difference between the observer and the observed and the world is strictly causally closed. As a consequence there can be no Heisenberg cut and no collapse of the wave function. This picture holds true in the many-worlds formalism. The Heisenberg cut is therefore inexistent with respect to the hypothetical third-person observers in the same way that for the infalling observer who is freely falling into the black hole the horizon is no special place in spacetime since she sees nothing special happening there (the equivalence principle). These global third-person observers, although they are hypothetical (since they are not observed directly in the world around us), their existence is fully logical (in contrast to Nietzsche's view).

However, there is another complementary perspective associated with the realistic local first-person observers of the Copenhagen interpretation and with respect to whom (i) there is indeed a fundamental distinction between the observer (classical system) and the observed (quantum system) and (ii) the collapse of the wave function is a genuine physical effect which is actually verified experimentally. See the quantum Zeno effect [33, 34].

It can be argued in this case, by following von Neumann, that the most natural placement of the Heisenberg cut is the interface between the consciousness of the observer, considered as a different substance, and the physical brain which is a part of the material substance (see for example [35] and references therein). The degrees of freedom associated with consciousness are thus constituted of a mental substance (a kind of dark energy which underlies mental phenomena) which is inaccessible to the first-person observers since the only degrees of freedom which can be physically accessed in the usual way by any observer (local or global) must be part of the material substance. Thus, these degrees of freedom are inaccessible since they lie effectively behind a horizon (the Heisenberg cut) and therefore they are analogous to the degrees of freedom found behind the horizon of a black hole which are inaccessible to the asymptotic observer.

Physicalism, i.e. the view that everything is reduced to matter and that the world is causally closed, underlies all of physics including quantum mechanics. But this idea of physicalism (or something near enough as advocated for from a physicalist point of view in [19]) is already challenged in the philosophy of mind by many philosophers such as Nagel [36] and Chalmers [11].

For example, according to Chalmers in his theory of 'naturalistic dualism', both the usual degrees of freedom of matter and the degrees of freedom associated with consciousness are equally fundamental and cannot be reduced to one another. In this case an appropriate ontology is really a double-aspect theory (such as the neutral monism of Spinoza) in which the material physical substance and the mental psychological substance are two aspects of a more neutral and a more fundamental substance which could perhaps be information, i.e. the 'it from bit' of Wheeler [37] (see also Sayre [38]). This is therefore a non-reductive theory which contains, in addition to the usual physical laws, psychophysical laws (or further facts) which determine how the mental arises from the physical. More precisely, subjective experience or qualitative consciousness (or qualia) may arise from (caused by) the physical but it is not entailed by the physical, i.e. it is ontologically distinct and hence it cannot be reduced to it.

Qualia is the so-called hard problem of consciousness, which cannot be explained functionally in contrast with the easy problems of consciousness which, according to Chalmers can, be explained functionally or neuronally (lower-level) or cognitively (higher level). This phenomenal consciousness (also called qualia) is what is associated with the subjective first-person experiences (like those of the Copenhagen observers) whereas functional consciousness or awareness is what is associated with the objective third-person experiences (like those of the many-world observers).

The role of the third-person observers of the many-worlds formalism also seems to be very similar to the role of the philosophical zombies in Chalmer's theory, in the sense that they are both logically possible although they are not directly seen in nature, which means in particular that qualia, sentience, thought, value and perhaps even intentionality and the like (which involve a first-person experience) require further facts for their explanation. Thus, there exists an explanatory gap as we go from the objective third-person level to the subjective first-person level of consciousness which can only be accounted for by an effective theory of naturalistic dualism which should admit neutral monism as an underlying fundamental theory.

Physicalism is then challenged by the subjective experiences (which are tied to the existence of qualia such as color) of first-person conscious observers existing in this world. However, physicalism also seems to be challenged by the experiences of these first-person observers, considered within the Copenhagen interpretation, who can observe the collapse of the wave function (a fact tied to the conscious experience of time and compatibilist free will).

Indeed, in the same way that neuronal interactions in the brain (themselves due to the interaction of the brain with the physical system mediated by the environment) give rise to the subjective experience of color (the redness of red for example) which is ontologically distinct from the brain itself, the interactions of the ontologically distinct degrees of freedom associated with the consciousness of the observer with the observed physical system (mediated through the brain and then the environment) give rise to the so-called collapse of the wave function (dead or alive cat). It appears therefore that the collapse as a (fifth) force works in the opposite direction of the psychological force producing qualia.

8.6 Quantum dualism, Wigner's friend and solipsism

By accepting consciousness as an ontologically different substance with independent degrees of freedom in addition to the usual physical particles and fields of the material substance, we reach the inescapable conclusion that the Heisenberg cut between the classical conscious observer and the quantum observed system should be necessarily and naturally placed at the demarcation line between the physical brain and the non-physical mind. At this point, the Copenhagen interpretation is essentially and effectively reduced to its logical limit which is the Wigner–von Neumann interpretation, also called 'the mind causes collapse' interpretation.

The complementarity between the Copenhagen interpretation and the many-worlds formalism implies therefore a distinct dualism (quantum dualism) between consciousness (here in the form of the collapse of the wave function which is caused by the measurement of first-person observers) and matter (here in the form of the unitary branching of parallel worlds and their linear coherent superposition as seen by third-person observers).

This is different from Cartesian dualism, shares some common features with naturalistic dualism but really it should be viewed as some form of neutral monism in which reality is constituted neither of consciousness nor of matter but of a more neutral substance with consciousness and matter being two different facets of it (not necessarily the only ones). The quantum dualism used by the first-person observers existing in this world should also be seen as complementary to the physicalism used by the hypothetical third-person observers of the many-worlds formalism (who are akin to the philosophical zombies of naturalistic dualism).

However, in naturalistic dualism the qualitative consciousness states (color and qualia in general) although ontologically independent of the physical, are produced by a physical system (the brain) but in quantum dualism it is the classical physical states that are produced (brought from potentiality to actuality) by the ontologically independent consciousness of the first-person observers (through the collapse) which leads directly to the (far-fetched in the view of many) proposition of solipsism. The complementarity relation between the Copenhagen interpretation and the many-worlds formalism is thus translated into a complementarity relation between solipsism and the many-worlds formalism, which should be viewed as dual properties.

Indeed, if by Bell's theorem the classical pointer or preferred states of a quantum system do not actually exist (but only exist in potentiality) until a quantum measurement is completed, and if by the Wigner–von Neumann interpretation a quantum measurement requires for its completion the free action of a conscious observer, then it follows naturally and logically that the degrees of freedom of the consciousness of the observer are responsible for bringing the classical states from potentiality into actuality, which we see as collapse.

A somewhat explicit mechanism for the action of consciousness on matter can be given in analogy with the many-minds interpretation [13] as follows.

The hypothetical third-person observer (Wigner) of the many-worlds formalism sees directly the coherent linear superposition (8.7), i.e. the superposition of the decayed-atom/dead-cat/sad-observer and undecayed-atom/living-cat/happy-observer (where the 'observer' here refers only to the physical brain states of the first-person observer). By assuming an infinity of classical mind states, associated with the non-physical degrees of freedom of consciousness, evolving stochastically in time with a probability given by Born's rule, then it is observed that each mind state evolves in a stochastic way to either being attached to the decayed-atom/dead-cat/sad-observer branch or to the undecayed-atom/alive-cat/happy-observer branch. In other words, the first-person observer (Schrödinger) in the superposition (8.7) becomes fully conscious of the content of the classical state to which it is attached, which appears therefore, from his perspective, as a collapse. This is the sense in which 'consciousness causes collapse', which is ultimately due to the fact that the classical mind or consciousness and the quantum matter both obey Born's rule, which indicates that the underlying fundamental neutral substance also obeys Born's rule. Also, it is not difficult to appreciate that this form of mental causation is nothing other than the Leibnizian psychophysical parallelism [14].

Thus, from the perspective of the subjective experiences of the first-person observer it is the mental degrees of freedom that guide the time evolution of the physical (the solipsism of the Wigner–von Neumann interpretation), whereas from the fact that quantum mechanics is more fundamental than classical mechanics it is seen that it is the physical degrees of freedom that guide the time evolution of the mental (third-person observer of the many-worlds). This duality between the two descriptions is of course due to the fact that both the classical mental and the quantum physical obey Born's rule in this scheme. But is quantum mechanics really more fundamental than classical mechanics or are they really two independent descriptions of two ontologically distinct substances of nature?

An independent argument for 'the mind causes collapse' interpretation is the largely underestimated or even underrated Wigner's friend experiment. Indeed, before Wigner performs his measurement the state $|\Psi\rangle$ of the joint system Schrödinger + cat + atom is given by the maximally entangled pre-measurement state (8.7). Wigner can surely ask his friend Schrödinger whether or not he saw a dead cat and then inspect the system cat + atom. The probabilities according to the Born rule are as follows:
- There is a probability 1/2 that Schrödinger will say 'yes' and the system from then on behaves as if it is in the state $|1\rangle$ of a dead cat.
- There is a probability 1/2 that the friend will say 'no' and the system from then on behaves as if it is in the state $|0\rangle$ of an alive cat.

In other words, it is for certain that the friend Schrödinger will say that he found a dead or alive cat, as the case may be, before Wigner asked him. This means in particular that in the reference frame (so to speak) of Schrödinger the state vector, even before Wigner's measurement, was already either $|1\rangle$ or $|0\rangle$ and not their linear combination, which is in gross contradiction to the quantum mechanical rule (8.7) verified experimentally to a great accuracy.

This is not to say that Schrödinger's position is less reasonable since quantum mechanics assumes him (in the reference frame of Wigner) to occupy the linear combination $|\Psi\rangle$ which implies in a clear sense, as Wigner puts it, 'that my friend was in a state of suspended animation before he answered my question' [22]. In other words, third-person observers (with respect to whom there is no collapse) really act as if they were philosophical zombies.

This experiment shows also among other things the non-tenability of objective collapse models (every measurement will produce a collapse for everybody) as opposed to the subjective-collapse models such as the Copenhagen interpretation (in which every observer is assigned a collapse in her own measurement only).

In summary, quantum mechanics enjoys four properties, which if taken at face value, makes the corresponding interpretation fall easy prey to solipsism (or subjective idealism). These are:
1. Quantum mechanics is naturally a dualistic theory in which the dual relation between observer and observed is lifted to a dual relation between consciousness and matter considered as two different aspects of a more fundamental and more neutral substance.
2. Quantum mechanics is strongly perspectival in character, i.e. the role of the observer is irreducible and objective reality is nothing less or more than an omni-perspective which is a total coherent sum of all the perspectives of all the observers existing in this world.
3. The world is not causally closed (taking into account the physical alone) if observers and consciousness cannot be eliminated. Indeed, the collapse of the wave function requires the action of a conscious observer. In other words, consciousness acts as a kind of 'dark energy' with causal influence on matter.

4. The (classical pointer) states of the physical system (or the world) exist only potentially and their existence becomes actualized or realized only when a measurement is performed on the physical system (or the world) in accordance with Bell's theorem. In other words, a transition from potential to actual existence occurs only when the wave function collapses. The collapse of the wave function acts therefore as a completely independent 'fifth force' in nature connecting the mental to the physical similar to the psychophysical force producing qualia which connects the physical to the mental.

As it turns out, most of the interpretations of quantum mechanics and most of the philosophy of (quantum) physics are in fact an attempt to explain away the above four straightforward properties of quantum mechanics. The only known exception to this rule is the Wigner–von Neumann interpretation (also called 'the mind causes collapse' interpretation) which can be defined precisely by the above four properties. These properties amount effectively to a solipsism which is a reality in which only the mind of the knowing subject or the observer is objectively real and everything else is only an appearance, i.e. an idea in the mind of the observer with no thing-in-itself behind it.

However, the form of solipsism suggested by property 4 is much weaker than this since it states that the classical states which are actualized by the act of observation are actually actualized by an objectively real psychophysical action of a conscious observer (since measurement requires a conscious observer in this scheme) and hence the action (not only the consciousness) of the observer is also real within this interpretation which should extend to the entirety of the psychophysical world.

8.7 Simulation and matrix hypotheses

By analogy with the simulation hypothesis [39] the first-person observers of the Copenhagen interpretation (who see the collapse of the wave function) play the role of the simulated beings populating the simulation, whereas the third-person observers of the many-worlds formalism (who are wholly unitary) play the role of the biological beings who are running the simulation.

Therefore, the simulated conscious first-person observers are like players in a giant virtual reality game and what they observe in the simulation is the rendering of the content of the simulated environment which appears to them as the collapse of the wave function when reality is finally experienced.

For quantum systems the simulated reality is not computed until the observer or the player seeks the experience or observation. This is simply due to the high cost of the calculations and to the limited resources available to the simulator. This is then the statement that the electron is not out there until observed (Bell's theorem). However, in classical systems the numerical calculation is low cost and therefore the computer performs the calculation well before the observation or rendering time. So the Moon is really out there even when we are not looking at it.

The above definition of physical reality as the connection between the rendering of the simulated environment to the player and the collapse of the wave function seen by the observer is based implicitly on the assumption of finite computation resources and requirement of low computational complexity [40]. Furthermore, if we are living in a simulated reality it is easy to imagine that the beings running the simulation have also computed the other parallel branches of the world. In fact this is only natural from the simulator's point of view. These beings are the third-person observers of the many-worlds formalism who observe directly coherent linear superpositions and thus the corresponding global structure of reality, not only the local one associated with Copenhagen.

The simulation hypothesis provides therefore a vivid metaphor or visualization of the laws of quantum mechanics. But it can also be considered as a genuine metaphysical theory in its own right. In other words, we are indeed likely to be among the simulated beings rather than among the biological ones. This provides therefore a powerful starting point for a new interpretation of quantum mechanics.

A generalization of the simulation hypothesis is the 'Matrix' hypothesis of David Chalmers [41] which is a modern formulation of the old philosophical riddle: 'the brain in a vat'. All these hypotheses lend themselves quite naturally to the quantum world.

8.8 The arrow of time and the Boltzmann–Schuetz hypothesis

The arrow of time is traced back to the thermodynamical arrow which in turn is traced back to the collapse postulate and observer-determinacy (Wigner–von Neumann) or observer-participancy (Copenhagen), and recalling Carter's words 'although the…situation [of the observer] is not necessarily central it is inevitably privileged to some extent', this privileged situation is traced back to the expansion of the Universe and the very special initial conditions at the Big Bang. In some sense space and its expansion are at the origin of the arrows of causation, consciousness and time.

The 'arrow of time' is a term due originally to Eddington [42] which refers to a paradox (identified originally by Boltzmann) between [43]:

- *A time-symmetric physics*: Indeed, all physical laws are symmetric under time reversal symmetry $T : t \longrightarrow -t$. The only exception is the second law of thermodynamics.
- *A time-asymmetric Nature*: Indeed, many natural processes are observed to be asymmetric in time, i.e. their corresponding time-reversed processes do not occur in Nature. Examples include (passage of time, thermodynamic asymmetry, radiation arrow, expansion of the Universe, causality, stream of consciousness). All these processes are associated with entropy increase and hence all of them are subsumed under the heading of the second law of thermodynamics.

Hence, 'arrow of time' is identified with the 'thermodynamical time-asymmetry' (because of the notion of entropy increase).

According to Price [44] there are three solutions (although he dismissed the third solution completely):

- *The causal-general approach*: This requires the postulation of an asymmetric dynamical (causal) law to explain the observed asymmetric behavior. But this approach will more likely also entail another asymmetry in the boundary condition. Hence, this is a two-asymmetry proposal. Examples: the H-theorem (molecular chaos assumption) due to Boltzmann himself and the collapse of the wave function [45].

- *The acausal-particular approach*: This is a one-asymmetry proposal which is the second view of Boltzmann in which the thermodynamical asymmetry is derived uniquely and solely from the initial conditions (hence it is acausal). In particular, a low-entropy initial state of the Universe guarantees the validity of the second law throughout the future. And the problem is thus reduced to cosmology and to the question why the state of the Universe was so special in the beginning (hence this approach is particular).
- *Boltzmann–Schuetz hypothesis*: The third solution is the Boltzmann–Schuetz hypothesis which is an idea due originally to Schuetz [46].
    - This is a no-asymmetry proposal, better known as the *acausal-anthropic approach*.
    - This is a serious contender but requires the (i) multiverse of quantum cosmology, (ii) the anthropic principle and (iii) the simulation hypothesis, as argued in [47].
    - These requirements are considered drawbacks of this alternative by [44] (and by the majority of physicists) but viewed positively in [47].
    - Indeed, the acausal-anthropic approach should be thought of as the fundamental theory of time from the perspective of the multiverse as a whole whereas the causal-general and acausal-particular approaches are effective descriptions from the perspectives of the conscious observer and the Universe, respectively.
    - Also, it is seen that the Boltzmann–Schuetz hypothesis, the anthropic principle and the simulation hypothesis are mutually reinforcing propositions.

The Boltzmann–Schuetz hypothesis (see figure 8.1) consists of the following assumptions:
- The universe is infinite in extent and is of an infinite age. This is the multiverse which is in the state of thermal equilibrium most of its time and thus has a constant entropy (heat death).
- Hence, the probability of producing low-entropy fluctuations which resemble our world becomes no longer small (although these fluctuations are still very rare). Our universe is one of these low-entropy fluctuations.
- The mind (intelligent observers) can only exist in these low-entropy fluctuations which, although they are very rare, are the only regions which can produce and then sustain life and consciousness through an entropy gradient and as a consequence can contain observers that can actually observe this universe. This is the anthropic principle.
- The Boltzmann measure shows then that smaller 'spontaneous' fluctuations are much more probable than the low-entropy 'real' fluctuation of our world. This leads immediately to the result that our universe is in fact simulated.

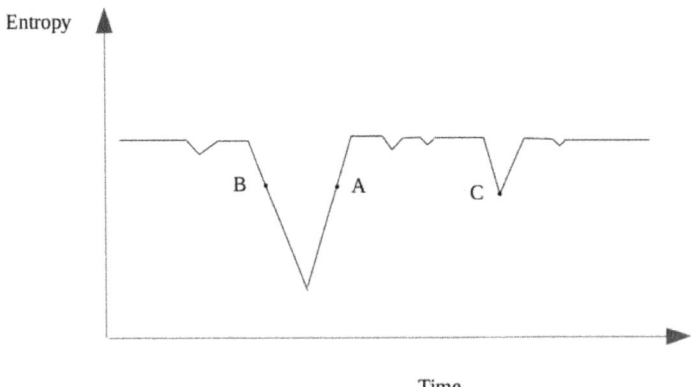

Figure 8.1. The Boltzmann–Schuetz hypothesis (are we in C or A?).

- The state of the Universe will evolve toward the equilibrium. We are thus found at points A where an entropy gradient exists in which entropy increases with time and time flows uphill from the past to the future.
- But other intelligent observers can exist at points B where the entropy gradient corresponds to a decrease in entropy with time flowing from past to future downhill. Thus, the arrow of time is certainly observable (objective) with respect to these local observables and also with respect to the Universe itself but vanishes completely at the much larger scales of the multiverse.
- However, it is much more probable that we actually exist at point C, which have the same entropy as points A and B, because it belongs to a much smaller fluctuation with a much higher entropy. Indeed, probability is proportional to the exponential of entropy and thus the 'spontaneous' fluctuation which contains C is much more probable than the 'real' low-entropy fluctuation of our world which contains A.

  As Price puts it 'if our own region has a past of even lower entropy it is much more improbable than it needs to be given its present entropy' [43].

- Thus, it is more probable that we are at C not at A or B (although they all have the same information content because they have the same entropy) and hence all our memories, records and histories are false or fake or illusionary.

  This point is due originally to Weizsäcker in 1939 who concludes this argument by stating that 'the most probable situation by far would be that the present moment represents the entropy minimum, while the past, which we infer from the available documents, is an illusion' [48].

- These false or fake or illusionary records and documents are best understood as simulated histories and memories. Thus, the Boltzmann–Schuetz hypothesis leads immediately to the fact that our universe is in fact simulated.

  However, this argument relies on the fact that simulation is cheaper in entropy terms than reality which can also be disputed [47]. Indeed, it can be argued within the many-worlds formalism of quantum mechanics and decoherence that the information content of the Universe as a whole is close to zero and that the observed complexity in Nature is only apparent seen from the perspective of the intelligent observers inhabiting this Universe [49]. Simulating such a reality with all its apparent complexity is certainly more expensive than this reality itself.

8.9 Time, free will and the quantum past

The fact that there are two types of change in time within the Copenhagen and Wigner–von Neumann interpretations means in particular that there are two fundamental measures of time which are not necessarily the same. The unitary evolution of the Schrödinger equation corresponds precisely to the objective physical time. Whereas the collapse of the wave function in the measurement process (as seen by first-person observers such as Schrödinger) corresponds to the subjective psychological time experienced by consciousness. It is the psychological time that has an arrow which is perspectival in character. These two times are not necessarily identical and they are expected to be unified (resulting in truly quantum time) within the framework of a new psychophysical theory of quantum mechanics (quantum gravity) in which a fine-grained unitary time evolution must replace the gross-grained Schrödinger time evolution which terminates dynamically (leaving the collapse in its place) each time a measurement is performed by a conscious observer.

The existence of two measures of time: (i) the objective physical time (unitarity) and (ii) the subjective psychological time (consciousness and collapse of the wave function) could be at the root cause of the problem of temporal non-locality in neuroscience discovered in experiments by Libet in 1983 [50]. It was discovered in these experiments that the readiness potential in the brain (which precedes voluntary movement) begins to arise in the brain before the conscious decision to move. By thinking about the conscious decision to move as marking the subjective psychological time associated with the collapse of the wave function, whereas thinking about the rising of the readiness potential as marking the objective physical time we can see that the discrepancy found by Libet can be interpreted as a slight failure of synchronization between the physical and the mental, i.e. the psychophysical parallelism is not exact. This is usually interpreted as indicating a compatibilistic free will.

Finally, it is observed that the nature of time and the nature of consciousness, from various quite distinct perspectives, are unified within quantum mechanics in the theory of observation. The most important properties of time is its flow and its direction, both of which are captured in the distinction perceived by conscious observes between past (retrodiction and memory) and future (prediction and free will). Thus, the unification between past, the present and future, i.e. the so-called problem of quantum past, should allow us an unprecedented window into these issues. See for example [51].

8.10 The mind/body problem, synchronicity and Bell's theorem

In this final section we discuss the mind/body problem and its similarity to Bell's theorem and the concept of synchronicity and its contrast with causality. In particular, synchronicity can be understood as the non-causal link relating observer and observed, i.e. the collapse postulate can be understood as Jung's and Pauli's synchronicity which is quite different from the parallelism of Leibniz.

8.10.1 The mind/body problem
Let us first start with the old philosophical paradox known as the mind/body problem. This can be summarized in the following points.
- For Descartes the mind is the ghost in the machine (body).
- Cartesian dualism is a substance dualism which states that there exists two substances: Matter and Mind. Each substance comes with attributes (space; thought) and modes (location, shape, size, weight; emotion, belief, desire).
- Descartes: All mental phenomena are conscious.
- Epistemology: Consciousness is private and subjective whereas matter is public and objective.
- Interactionism: Matter acts on mind and mind acts on matter (common sense view). But how can this be possible with the Cartesian metaphysical distance between the two?
- The mind/body problem consists of the mutual contradiction between the following three hypotheses: (i) dualism, (ii) causal closure and (iii) mental causation.
- Quantum mechanics seems to challenge causal closure (the observer is participatory not a spectator). Similarly, Cartesian dualism adopts dualism and mental causation and rejects causal closure.
- There are many other solutions, for example functionalism, idealism and double-aspect theory (neutral monism), that reject dualism whereas occasionalism (Al-Ghazal) and parallelism (Leibniz) reject mental causation.
- Another solution: Chalmers' naturalistic dualism in which reality is constituted out of the elementary particles and the elementary moments of consciousness.
- The mind is a triangle with three summits (experience, attitude and action) related by five dimensions (observation, access, expression, direction, theory). Emotions are in the middle of the triangle.
- Language, thought and belief are found in (attitude) whereas consciousness and perception are found in (experience). The difference between consciousness and perception is that perception is functional.
- (Experience) is challenged by the explanatory gap, (thought) is challenged by Godel's theorem whereas (language) is challenged by emergence.
- (Action) represents agency and free will. This is challenged by causation and determinism.

8.10.2 Synchronicity
- Synchronicity is a concept due to Carl Jung and Wolfgang Pauli [52, 53] (see also [54, 55]) which is meant to be 'an acausal connecting principle', 'meaningful coincidence' and 'meaning-correspondence'.

- The two concepts of synchronicity and archetypes are the most important concepts for us here. They are the mental analogues of causality and particles.
- Synchronicity and causality should be unified in a single category of cognition in the same way that space and time are unified in a spacetime.
- Causality comes, according to Aristotle, in the form of four causes (final, formal, material and efficient). Teleology (final) might be related to synchronicity. Only material and efficient causes are usually accepted (since Hume).
- Causality is rooted in forces whereas synchronicity is rooted in acausal influences and constraints such as the Pauli exclusion principle. Another example is Bell's correlations.
- Causality rules the physical world, whereas occasionalism rules the mental world and synchronicity provides the link.
- The link between matter and mind is in quantum mechanics (observation and the collapse postulate). The collapse of the wave function and the observer–observed entanglement (or more precisely the consciousness–world entanglement) is a synchronicity event.
- Causality is expressed by forces between particles and the forces and particles are underlied by symmetries. Similarly, synchronicity is expressed by meaningful coincidences between acausal events which are underlied by archetypes (here archetype play a dual role of particles/forces and symmetries). Most coincidences between events are actually causal but some cannot be explained by causality and furthermore carry meaning. Archetypes are particles of the mental which interact via the forces given by synchronicities.
- Causality is constant connection through effect whereas synchronicity is inconstant connection through contingency, equivalence and meaning.
- Jung: The occurrence of a meaningful coincidence is of three forms:
    - The coincidence of a certain psychic content with a corresponding objective process which is perceived to take place simultaneously.
    - The coincidence of a subjective psychic state with a phantasm (dream or vision) which later turns out to be a more or less faithful reflection of a 'synchronistic,' objective event that took place more or less simultaneously, but at a distance.
    - The same, except that the event perceived takes place in the future and is represented in the present only by a phantasm that corresponds to it.

8.10.3 The world-from-decoherence and the world-in-consciousness

- We define the three terms (reality), (world) and (psyche).
- Reality is everything that exists which is described by quantum mechanics.
- The principle of linear superposition is the hallmark of all quantum mechanical phenomena. This is the property of many-worlds. In other words, reality is the many-worlds.
- Thus, the electron in the hydrogen atom exists in a linear combination of say the ground state and the first excited state. A quantum bit is found in a linear combination of spin up and spin down, and Schrödinger's cat is found in a

linear combination of dead and alive. Thus, reality is a linear superposition of all possibilities.
- Reality is thus everything that exists in a linear superposition.
- Now, the world is the macroscopic classical world of all phenomena which we think we are inhabiting.
- Reality under decoherence (the decoherence of the various possibilities found in the linear superposition) will become the macroscopic classical world. Decoherence is a causal phenomenon.
- But quantum mechanics contains also the phenomena of observation or measurement. The observer will see either a live cat or a dead cat but never a linear combination. The third-person observers who can see those linear superpositions exist in logic but are absent in nature.
- Thus reality under the effect of observation will also become a classical world, but this classical world is perspectival and contextual.
- Thus the world-from-decoherence (objective) and the world-in-consciousness (perspectival) are really two different worlds and not necessarily one. This is the source of confusion in many topics of the philosophy of consciousness and the philosophy of quantum mechanics.
- The world-from-decoherence is the material and sensible world with no place in it for consciousness (a horizontal line).
- In contrast the world-in-consciousness is the world of the psyche. The state of the cat after measurement is actually the tensor products: (alive cat $\otimes$ happy observer) and (dead cat $\otimes$ sad observer). The world is classical from the perspective of this observer. The act of measurement also produced psychological states.
- The act of measurement produces a world in the mind of the observer (Berkely) which is different from the world-out-there which is obtained from decoherence. This world is the psyche (a vertical line).
- Classical mechanics is an approximation of quantum mechanics only if we think about the world-from-decoherence. However, from the perspective of the world-in-consciousness classical mechanics is the fundamental theory of cognition (the complementarity principle according to Bohr is the only way to understand nature causally in spacetime).
- Decoherence is causal but the collapse of the wave function and the entanglement between observer and observed during the measurement is an 'acausal meaningful connection', i.e. a synchronicity. This is a coincidence between an objective physical phenomenon and a subjective conscious state.

8.10.4 Bell's theorem revisited

Bell's theorem relies on three hypotheses:
- Classical realism.
- Local causality.
- Free will.

Nature and quantum mechanics both violate Bell's theorem. In other words, not everything is logical.

However, we observed a stark resemblance between the hypotheses of Bell's theorem and the hypotheses underlying the mind/body problem:
- Dualism.
- Causal closure.
- Mental causation.

It seems that both Nature and quantum mechanics violate also these hypotheses. In other words, not everything is physical. In particular, quantum mechanics seems to suggest the alternative quantum hypotheses:
- Quantum dualism.
- Causal/synchronistic closure.
- Perspectivism.

A solution to the mind/body problem starting from these quantum hypotheses is very possible. See figure 8.2.

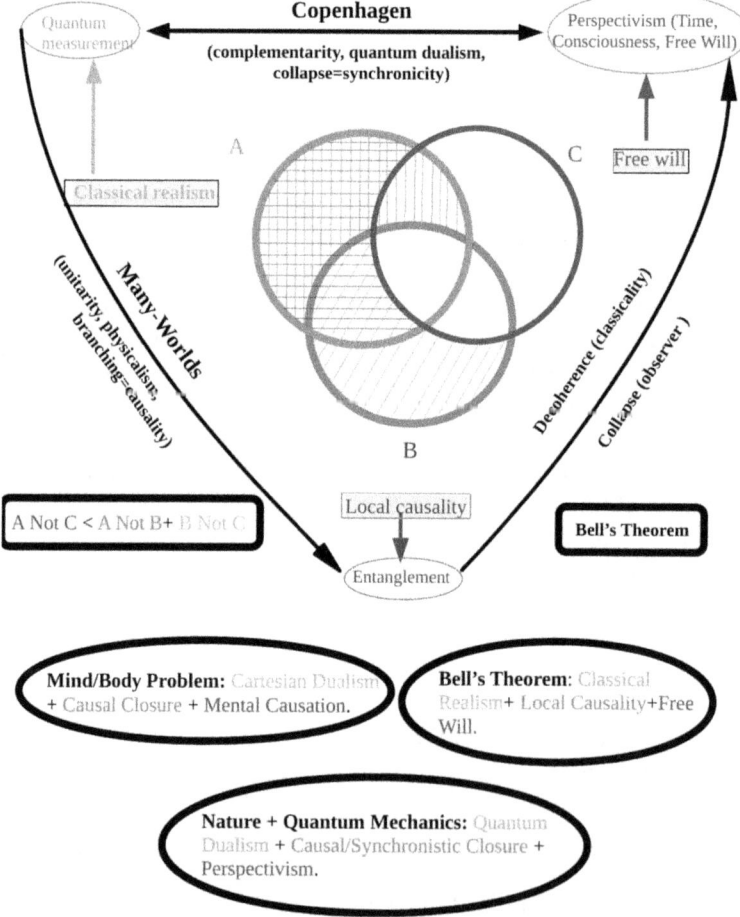

Figure 8.2. Bell's theorem and the mind/body problem.

8.11 Conclusion

The Wigner–von Neumann interpretation of quantum mechanics was presented in this chapter. It is argued that this interpretation, if taken as an effective theory, will involve four interwoven elements which are: (i) quantum dualism, (ii) perspectivism as an extended complementarity principle, (iii) psychophysical causal closure and (iv) solipsism. However, as a fundamental theory the Wigner–von Neumann interpretation depicts a reality of neutral monism which is complementary to the physicalism of the many-worlds formalism.

Indeed, within this scheme we have on one hand the fact that in the single-world the interactions of the ontologically distinct degrees of freedom associated with the consciousness of the classical observer with the physical degrees of freedom of the quantum physical system give rise to the collapse of the wave function, whereas on the other hand we have the complementarity relation between the Copenhagen interpretation and the many-worlds formalism. This means in particular that the degrees of freedom associated with the consciousness of the first-person observer in the single-world are dual to the purely physical degrees of freedom associated with the many-worlds formalism. As a consequence the quantum dualism employed by the first-person observers is a complementary description to the physicalism employed by the third-person observers.

The property of solipsism, which is seen as dual to the property of many-worlds, can be given a much weaker import by appealing to the many-minds interpretation and the underlying neutral monism. In this chapter we have also discussed, among many other things, two fundamental topics of interest to the philosophy of quantum physics which are (i) the Boltzmann–Schuetz hypothesis and (ii) the mind/body problem.

References

[1] Dirac P 1938 A new basis for cosmology *Proc. R. Soc. Lond.* A **165** 199–208
[2] Dicke R H 1961 Dirac's cosmology and Mach's principle *Nature* **192** 440
[3] Dicke R H 1957 Gravitation without a principle of equivalence *Rev. Mod. Phys.* **29** 363–76
[4] Carter B 1974 Large number coincidences and the anthropic principle in cosmology *IAU Symp.* **63** 291
[5] Barrow J D and Tipler F G 1986 *The Anthropic Cosmological Principle* (Oxford: Oxford University Press)
[6] Schombert J *Cosmology* (Lecture notes) (http://abyss.uoregon.edu/js/cosmo/)
[7] von Neumann J 1955 *Mathematical Foundations of Quantum Mechanics* 1st edn (Princeton, NJ: Princeton University Press)
[8] Wigner E P 1967 *Symmetries and Reflections* (Bloomington, IN: Indiana University Press)
[9] Edwards D A 1979 The mathematical foundations of quantum mechanics *Synthese* **42** 1
[10] Birkhoff G and von Neumann J 1936 The logic of quantum mechanics *Ann. Math.* **37** 823–43
[11] Chalmers D 2007 The hard problem of consciousness *The Blackwell Companion to Consciousness* ed M Velmans and S Schneider (Oxford: Blackwell)
Chalmers D 2007 Naturalistic dualism *The Blackwell Companion to Consciousness* ed M Velmans and S Schneider (Oxford: Blackwell)
Chalmers D 2010 *The Character of Consciousness* (Oxford: Oxford University Press)

[12] Bohr N 1963 *Essays 1958–62 on Atomic Physics and Human Knowledge* (New York: Wiley)
Bohr N 1949 Discussions with Einstein on epistemological problems in atomic physics *Albert Einstein: Philosopher-Scientist* (Cambridge: Cambridge University Press)

[13] Albert D and Lower B 1988 Interpreting the many worlds interpretation *Synthese* **77** 195–213

[14] Kulstad M and Carlin L 2020 Leibniz's philosophy of mind *The Stanford Encyclopedia of Philosophy* **Winter 2013** https://plato.stanford.edu/entries/leibniz-mind/

[15] Bell J S 1964 On the Einstein–Podolsky–Rosen paradox *Physics* **1** 195

[16] Bell J S 1966 On the problem of hidden variables in quantum mechanics *Rev. Mod. Phys.* **38** 447

[17] Descartes R 1642 *Meditationes de Prima Philosophia* 2nd edn (Whitefish, MT: Kessinger Publishing LLC)

[18] de Spinoza B 1992 *Ethics* ed A Boyle (London: Everyman's Library)

[19] Kim J 2005 *Physicalism or Something Near Enough* (Princeton, NJ: Princeton University Press)

[20] Heisenberg W 2011 Is a deterministic completion of quantum mechanics possible? ed E Crull and G Bacciagaluppi https://halshs.archives-ouvertes.fr/halshs-00996315

[21] Trimmer J D 1980 The present situation in quantum mechanics: a translation of Schrödinger's 'cat paradox' paper *Proc. Am. Phil. Soc.* **124** 323–38

[22] Wigner E 2001 Remarks on the min d body question *Philosophical Reflections and Syntheses. The Collected Works of Eugene Paul Wigner. Part B: Historical, Philosophical, and Socio-Political Papers* ed J Mehra vol B/6 (Berlin: Springer)

[23] Zeh H D 1970 On the interpretation of measurement in quantum theory *Found. Phys.* **1** 69–76

[24] Everett H 1957 Relative state formulation of quantum mechanics *Rev. Mod. Phys.* **29** 454–62

[25] Everett H, Wheeler J A, DeWitt B S, Cooper L N, Van Vechten D and Graham N 1973 *The Many-Worlds Interpretation of Quantum Mechanics* Princeton Series in Physics ed B DeWitt and N Graham (Princeton, NJ: Princeton University Press)

[26] Tegmark M 1998 The interpretation of quantum mechanics: many worlds or many words? *Fortsch. Phys.* **46** 855–62

[27] Susskind L 2016 Copenhagen vs Everett, teleportation, and ER=EPR *Fortsch. Phys.* **64** 551
Susskind L 2016 ER=EPR, GHZ, and the consistency of quantum measurements *Fortsch. Phys.* **64** 72

[28] Nietzsche F W, Kaufmann W A and Hollingdale R J 1964 *The Will to Power* (New York: Vintage Books) §481 p 267
Nietzsche F W and Kaufmann W A 1974 *The Gay Science; With a Prelude in Rhymes and an Appendix of Songs* (New York: Random House) §374 p 336
Nietzsche F 1998 *On the Genealogy of Morality: A Polemic* ed M Clarke and A J Swenswen (Indianapolis, IN: Hackett)

[29] Anderson R L 1998 Truth and objectivity in perspectivism *Synthese* **115** 1–32

[30] Gleason A M 1957 Measures on the closed subspaces of a Hilbert space *J. Math. Mech.* **6** 885–93

[31] Kochen S and Specker E 1967 The problem of hidden variables in quantum mechanics *J. Math. Mech.* **17** 59–87

[32] Einstein A, Podolsky B and Rosen N 1935 Can quantum mechanical description of physical reality be considered complete? *Phys. Rev.* **47** 777–80

[33] Misra B and Sudarshan E C G 1977 The Zeno's paradox in quantum theory *J. Math. Phys.* **18** 756

[34] Patil Y S, Chakram S and Vengalattore M 2015 Quantum control by imaging: the Zeno effect in an ultracold lattice gas *Phys. Rev. Lett.* **115** 140402
[35] Stapp H 2007 Quantum mechanical theories of consciousness *The Blackwell Companion to Consciousness* ed M Velmans and S Schneider (Oxford: Blackwell)
Stapp H P 2009 *Mind, Matter and Quantum Mechanics The Frontiers Collection* ed M Velmans and S Schneider (Berlin: Springer)
[36] Nagel T 1974 What is it like to be a bat *Phil. Rev.* **83** 435–50
[37] Wheeler J A 1990 Information, physics, quantum: the search for links *Complexity, Entropy, and the Physics of Information* ed W H Zurek (Redwood City, CA: Addison-Wesley)
[38] Sayre K M 1976 *Cybernetics and the Philosophy of Mind* (Notre Dame: University of Notre Dame Press)
[39] Bostrom N 2003 Are you living in a computer simulation *Phil. Q.* **53** 243–55
[40] Campbell T, Owhadi H, Sauvageau J and Watkinson D 2017 On testing the simulation theory arXiv:1703.00058 [quant-ph]
[41] Chalmers D 2005 The matrix as metaphysics *Philosophers Explore the Matrix* ed C Grau (Oxford: Oxford University Press)
[42] Eddington A 1928 *The Nature of the Physical World* (Cambridge: Cambridge University Press)
[43] Price H 2002 Time's arrow and Eddington's challenge *Br. J. Phil. Sci.* **53** 83
[44] Price H 2010 *Boltzmann's Time Bomb* Séminaire Poincaré XV Le Temps pp 115–40
[45] Albert D Z 2000 *Time and Chance* (Cambridge, MA: Harvard University Press)
[46] Boltzmann L 1895 On certain questions of the theory of gases *Nature* **51** 413
[47] Cirkovic M M 2003 The thermodynamical arrow of time: reinterpreting the Boltzmann–Schuetz argument *Found. Phys.* **33** 467
[48] von Weizsacker C 1939 Der zweite Hauptsatz und der Unterschied von Vergangenheit und Zukunft *Ann. Phys., Lpz.* **36** 275
von Weizsacker C 1980 *The Unity of Nature* (New York: Farrar Straus Giroux) (Engl. transl.)
[49] Tegmark M 1996 Does the Universe in fact contain almost no information? *Found. Phys. Lett.* **9** 25
[50] Libet B 2004 *Mind Time: The Temporal Factor in Consciousness* (Cambridge, MA: Harvard University Press)
[51] Hartle J B 1998 Quantum pasts and the utility of history *Phys. Scr.* **T76** 67
[52] Jung C and Pauli W 1955 *The Interpretation of Nature and Psyche* (New York: Pantheon Books)
[53] Jung C and Pauli W 1992 *Atom and Archetype: The Pauli/Jung Letters, 1932–1958* ed C A Meier (Princeton, NJ: Princeton University Press)
[54] Peat F D 1987 *Synchronicity: The Bridge Between Matter and Mind* (New York: Bantam Books)
[55] Peat F D 1999 Time, synchronicity and evolution *Kooperation mit der Evolution. Über das Zusammenspiel von Mensch und Kosmos* ed M Sachtleben (München: Diederichs, New Science)

Lightning Source UK Ltd.
Milton Keynes UK
UKHW030640111121
393781UK00004B/156